SpringerBriefs in Applied Sciences and Technology

Computational Intelligence

Series editor

Janusz Kacprzyk, Warsaw, Poland

About this Series

The series "Studies in Computational Intelligence" (SCI) publishes new developments and advances in the various areas of computational intelligence—quickly and with a high quality. The intent is to cover the theory, applications, and design methods of computational intelligence, as embedded in the fields of engineering, computer science, physics and life sciences, as well as the methodologies behind them. The series contains monographs, lecture notes and edited volumes in computational intelligence spanning the areas of neural networks, connectionist systems, genetic algorithms, evolutionary computation, artificial intelligence, cellular automata, self-organizing systems, soft computing, fuzzy systems, and hybrid intelligent systems. Of particular value to both the contributors and the readership are the short publication timeframe and the world-wide distribution, which enable both wide and rapid dissemination of research output.

More information about this series at http://www.springer.com/series/10618

Ranjit Biswas

Is 'Fuzzy Theory' an Appropriate Tool for Large Size Problems?

 Springer

Ranjit Biswas
Faculty of Engineering and Technology,
    Department of Computer Science
    and Engineering
Jamia Hamdard University
New Delhi
India

ISSN 2191-530X   ISSN 2191-5318   (electronic)
SpringerBriefs in Applied Sciences and Technology
ISBN 978-3-319-26717-3   ISBN 978-3-319-26718-0   (eBook)
DOI 10.1007/978-3-319-26718-0

Library of Congress Control Number: 2015956365

Printed on acid-free paper

This Springer imprint is published by SpringerNature
The registered company is Springer International Publishing AG Switzerland

Contents

Abstract

To the world's scientists and researchers, the most popular soft computing set theories are as follows: fuzzy set theory, intuitionistic fuzzy set theory (vague sets are nothing but intuitionistic fuzzy sets, as justified and reported by many authors), i–v fuzzy set theory, i–v intuitionistic fuzzy set theory, L-fuzzy set theory, type-2 fuzzy set theory, rough set theory, soft set theory, etc. In such theories, the value of $\mu(x)$ is proposed by the concerned decision maker by his best possible judgment. While a hungry tiger finds his food like one cow or one buffalo or one deer (or any other animal of his own list) in the forest, he decides a lot by his best possible judgment on a number of significant parameters before he starts to chase and also during the real-time period of his chasing. The decision makers are human being, animal/bird, or any living thing which has brain (we do not consider software or robots which have artificial intelligence). The most important (but yet an open unsolved problem) question in the theory of soft computing is: How does a cognition system of human/animal evaluate the membership value $\mu(x)$? This work unearths a ground-level reality about the 'Progress' of decision-making process in the human/animal cognition systems while evaluating the membership value $\mu(x)$ by proposing the **Theory of CIFS**. It is finally justified and concluded that 'fuzzy theory' may not be an appropriate tool to deal with large-sized problems, while in pursuance of excellent results. But at the end, two examples of 'decision-making problems' with solutions are presented, out of which one will show the dominance of the application potential of intuitionistic fuzzy set theory over fuzzy set theory and the other will show the converse, i.e., the dominance of the application potential of fuzzy set theory over intuitionistic fuzzy set theory in some cases.

Keywords CIFS · Atanassov trio functions · Atanassov initialization · Atanassov constraint · Atanassov trio bags · h-bag · m-bag · n-bag · Atanassov speed of convergence · Atanassov processing time (APT) · Atanassov absolute triangle · Decision trajectory curve · Atanassov point of convergence · Atanassov line-segment · Zadeh line-segment · Atanassov triangle · Intuitionistic μ-index ($i\mu i$) · Atanassov indicator · CESFM · Positive parameters · Negative parameters · CES-score (CS) · Fuzzy pocket machine · Punishment fuzzy set · CPS · PCS · NCS · CESFM-software

Is 'Fuzzy Theory' an Appropriate Tool for Large Size Problems?

1 Introduction

The most important question (but yet an open unsolved problem) in the theory of soft-computing is as mentioned below:

How Does the Cognition System of a Human or Animal (or of any living thing which has brain) Evaluate the Membership Value $\mu(x)$?

The work in this book is based on philosophical as well as logical views on the subject of decoding the 'progress' of decision making process in the Human/Animal cognition systems while evaluating the membership value $\mu(x)$ in a fuzzy set or in an intuitionistic fuzzy set or in any such soft computing set model or in a crisp set, developing a new theory called by **"Theory of CIFS"**. By 'cognition system' we shall mean throughout here the cognition system of a human being or of a living animal or of a bird or of any living thing which has brain (we do not consider robots or software which have artificial intelligence). At the end of this work one could realize the final outcome which is: "**at the ground reality of cognition system while evaluating a fuzzy membership value $\mu(x)$ or a membership value $\mu(x)$ corresponding to any soft computing set theory model or even crisp membership value (0 or 1), it is not possible to apply the fuzzy theory or any soft computing set theory or crisp theory in any decision making process without intuitionistic fuzzy theory**".

Quite naturally, three immediate questions arise in mind: (1) whether a crisp decision maker or a fuzzy decision maker or any soft computing decision maker needs to have knowledge about 'Intuitionistic Fuzzy Set Theory'? (2) Can they not make any decision if they are not aware/knowledgeable of intuitionistic fuzzy theory? and (3) What's about animals, birds or other living things who are also decision makers? For example, a hungry lion thinks and decides a lot before chasing a buffalo and also during the period of chasing a buffalo! Even in many situations they decide whether it is appropriate to chase, or even after chasing they

© The Author(s) 2016
R. Biswas, *Is 'Fuzzy Theory' an Appropriate Tool for Large Size Problems?*,
SpringerBriefs in Computational Intelligence, DOI 10.1007/978-3-319-26718-0_1

decide every moment whether to give up chasing or to continue chasing without any problem of his own security, etc.

This article deals mainly on these three questions initiating a new theory called by **Theory of CIFS (Cognitive Intuitionistic Fuzzy System)**. It is justified in length that a crisp decision maker or a fuzzy decision maker (or any soft decision maker) can not step forward without using intuitionistic fuzzy set (IFS) theory, but he does not necessarily need to have any knowledge of intuitionistic fuzzy theory. The permanent residence of IFS theory inside the brain (CPU) of every living thing is a hidden truth. The active performance of IFS theory is a permanent and hidden resident in the form of like a **'in-built system-software'** inside the kernel of the processor (i.e. at the lowest level) of every cognition system (be of human or of animal or of bird or of any living thing). This software like system automatically gets executed at the lowest level based upon the platform of intuitionistic fuzzy theory while evaluating any membership value $\mu(x)$ in a fuzzy set or in any soft-computing set or crisp set, the evaluation which always happens at higher level. Although this execution happens in a hidden way at the lowest level (like execution of a machine language program in CPU) but it continuously outputs to update the estimated value of $\mu(x)$ at higher level in the cognition system till some amount of time.

Let us focus our thoughts on: How the estimation process of the "membership value $\mu(x)$ of an element x to belong to a fuzzy set A or an IFS A" is initiated in the cognition system at time t = 0 and completed after time t = T (>0); How does in reality the 'progress' of decision making process for $\mu(x)$ actually happen inside the brain with respect to the variable 'time'.

In this work it is justified that the estimation process for $\mu(x)$ to construct a fuzzy set can never be executed in the human cognition system without intuitionistic fuzzy system. But for this, it does not require that a fuzzy decision maker or a crisp ordinary decision maker must be aware or knowledgeable about IFS Theory. Consider the case of a FORTRAN programmer who executes his program written by him in FORTRAN language corresponding to a given engineering problem. But for this, it does not require that the programmer must be aware or knowledgeable about machine language programming! The analogous fact is true for a fuzzy decision maker too, who estimates $\mu(x)$ using the domain of his fuzzy knowledge whereas at the lowest level inside his cognition system the exact execution happens under intuitionistic fuzzy systems only, which is justified in length in this work introducing the Theory of CIFS.

Consequently, it is true that although fuzzy sets are a special case of intuitionistic fuzzy sets, but the existing concept that "the intuitionistic fuzzy sets are higher order fuzzy sets" is **incorrect** (an example of similar **incorrect** concept can be imagined if somebody says that "fuzzy sets can be viewed as higher order crisp sets"!). Thus it is fact that the Theory of IFS is the most appropriate model for translation of imprecise objects while the fuzzy sets are 'lower order' or 'lower dimensional' intuitionistic fuzzy sets as special case. It is also rigorously justified here with examples that it may not be an appropriate decision to use fuzzy theory if the problem under study involves estimation of membership values for large number of elements.

Two examples of 'decision making problems' with solutions are presented out of which one example will show the dominance of the application potential of intuitionistic fuzzy set theory over fuzzy set theory, and the other will show the converse i.e. the dominance of the application potential of fuzzy set theory over intuitionistic fuzzy set theory in some cases (where decision makers are pre-selected and most intellectual in the context of the subject pertaining to the concerned problem).

2 Two Hidden Facts About Fuzzy Set Theory (and, About Any Soft Computing Set Theory)

The following two hypothesis are hidden facts in fuzzy computing or in any soft computing process, a careful analysis of which will unearth its correctness:-

Fact-1
A decision maker (intelligent agent) can never use or apply 'fuzzy theory' or any soft-computing set theory without intuitionistic fuzzy system.

Fact-2
The Fact-1 does not necessarily require that a fuzzy decision maker (or a crisp ordinary decision maker or a decision maker with any other soft theory models or a decision maker like animal/bird which has brain, etc.) must be aware or knowledgeable about IFS Theory!

For justifying these two facts, let us begin with a simple hypothetical example here (and at the end it is again justified by the **Proposition 8** here).

2.1 Understanding the 'Discovery of CIFS' by an Example

Suppose that there are 50 students in Class-V in Calcutta St. Paul's School. Let X be the set of all these 50 students, say $X = \{x_1, x_2, x_3, \ldots, x_{50}\}$. For execution of a statistical project, suppose that a decision maker (say, Mr. Smith) wants to consider the collection A of all the tall students of X. Suppose that by his "best possible judgment", Mr. Smith considers the fuzzy set A of the universe of discourse X:

$$A = \{(x_1, 0.80), (x_2, 0.65), (x_3, 0.93), \ldots, (x_{50}, 0.30)\}.$$

With no loss of generality, consider the student x_1 and the corresponding membership value $\mu(x_1)$ which is 0.80 here in A given by Mr. Smith by his "best possible judgment". It is obvious that the judgment process of Mr. Smith by which the student x_1 is given the membership value 0.80 in the fuzzy set A has not been finished in zero amount of processing time by the brain of Mr. Smith here. It may take an infinitesimally small amount of time Δt (of the order of 10^{-N} ns where N could be very very large) or it could be in few nano-seconds or in few micro-seconds or in few seconds/minutes or in hours or in days, etc. Suppose that the complete processing time taken by the decision maker Mr. Smith to come to his final judgment that "$\mu(x_1) = 0.80$" is T_1 (>0) unit of time.

Let us designate this time of processing for evaluating the membership value $\mu(x_1)$ as "**Atanassov Processing Time**" (APT) for the element x_1 corresponding to the decision maker Mr. Smith, and let us use the notation $APT_{Smith}(x_1) = T_1$ or in short the notation $APT(x_1) = T_1$. Thus, the value of $\mu(x_1)$ proposed by the decision maker is 0.80 for which the time-cost is T_1 (>0), and hence by fuzzy theory one can compute $\upsilon(x_1) = 1 - 0.80 = 0.20$ by doing an arithmetic just, without any further cost of time towards decision process.

I am here curious to know the continuous status of the brain (CPU) or the cognition system of the decision maker Mr. Smith during the various instants of time in the interval $[0, T_1]$. It is absolutely sure that at the starting time $t = 0$, evaluation of $\mu(x_1)$ was not complete and can not be complete, and hence its value can not be equal to 0.80 at time $t = 0$. It is also sure that at any instant of time $t < T_1$, i.e. in the right-open interval $[0, T_1)$, the complete evaluation of $\mu(x_1)$ was not over and hence the "under-process $\mu(x_1)$" ≤ 0.80 in the time interval $[0, T_1)$ because of the fact that it is a pre-mature stage of the cognition processor to output the final value of $\mu(x_1)$; and <u>finally</u> at time $t = T_1$, the mercury of "under-process $\mu(x_1)$" stops permanently at 0.80 inside the cognition system of Mr. Smith. Since value of "under-process $\mu(x_1)$" is sometimes ≤ 0.80 and then sometimes equal to 0.80, it is thus fact that "under-process $\mu(x_1)$" is a <u>continuous non-decreasing function</u> of 'time' t during the period of evaluation in the cognition system, until it become fully-matured being frozen at a constant value 0.80 at time T_1.

Let us designate this function by $m_{x_1}(t)$ or by $m(x_1, t)$ whose life period domain is the closed interval $[0, T_1]$ of time; i.e. the function $m(x_1, t)$ starts growing from time $t = 0$ and continues to grow (at least it does not diminish) till the time $t = T_1$ when it achieves the final functional value 0.80. This final value 0.80 is what we call the 'membership value' $\mu(x_1)$ in the fuzzy set A of X proposed by the concerned decision maker by his best possible judgment for the element x_1 of X. Thus $0 \leq m(x_1, t) \leq \mu(x_1)$, where $\mu(x_1)$ is a constant value (here it is 0.80) but $m(x_1, t)$ updates itself with the continuous progress of time starting from $t = 0$ till $t = T$.

Henceforth, it is obvious to us that the pre-matured values of the function $m(x_1, t)$ during the continuous progress of decision process are not to be assumed to be the 'membership value' $\mu(x_1)$, but the final value of $m(x_1, t)$ only. And hence in general for any element x of X, the $\mu(x)$ is a constant value, which is the maximum value of the function $m(x, t)$ with respect to its argument variable t (corresponding to a fixed decision maker).

But the basic question at this point is:

"At different rising instants of time t in $[0, T_1]$ starting from $t = 0$, what's about the growth and other properties of the function $m(x_1, t)$ in the cognition system of the decision maker, corresponding to the element x_1"?

We unearth the hypothesis that: at different instants of time t in $[0, T_1]$ starting from t = 0, the decision process runs under the "**Cognitive Intuitionistic Fuzzy System (CIFS)**" like machine language programming i.e. like the execution of a .EXE file of a source file. This remains hidden inside the kernel (at the lowest level) and un-communicated to the outer layer of the brain of the decision maker. At time $t = T_1$, the decision process completes its execution and yields the data 0.80 as the membership value of the object x_1 which is communicated all around at higher level peripherals, and finally exposed by the decision maker to the outer world of his brain.

3   Cognitive Intuitionistic Fuzzy System (CIFS)

In this section the details theory of the "Cognitive Intuitionistic Fuzzy System (CIFS)" is introduced. In the previous example it is justified that for a given decision maker and corresponding to an element x_1, the membership value $\mu(x_1)$ in a fuzzy set (or in an intuitionistic fuzzy set or in any soft computing set) is a constant value being the final value (maximum value) of a function $m(x_1, t)$. Consequently, the following trio functions corresponding to a given element x of the universe of discourse X is proposed.

3.1   Atanassov Trio Functions and Atanassov Constraint

Let R^* be the set of all non-negative real numbers. Consider a pre-fixed fuzzy decision maker. For any given element x of the set X to belong to the fuzzy set A of X, the membership value $\mu(x)$ is the final output of a hidden "cognitive intuitionistic fuzzy system" in the brain of the fuzzy decision maker where the following three functions are co-active:

(i) **h(x, t)** called by 'Hesitation Function' whose domain is $X \times R^*$ and range is [0, 1]. For a fixed element x of the set X, h(x, t) is a non-increasing continuous function of time t.

(ii) **m(x, t)** called by 'Membership Function' whose domain is $X \times R^*$ and range is [0, 1]. For a fixed element x of the set X, m(x, t) is a non-decreasing continuous function of time t.

(iii) **n(x, t)** called by 'Non-membership Function' whose domain is $X \times R^*$ and range is [0, 1]. For a fixed element x of the set X, n(x, t) is a non-decreasing continuous function of time t.

These three functions $\langle m(x, t), n(x, t), h(x, t) \rangle$ are called **Atanassov Trio Functions (AT functions)**.

These functions are subject to the constraint:

$$h(x, \ t) + m(x, \ t) + n(x, \ t) = 1 \quad \text{for any time t.} \ (\textbf{Atanassov Constraint}).$$

Each of the AT functions is to be basically treated as a function of time t, because it is always considered once the element x be picked up for evaluation by the decision maker. It is also important to note that the three trio functions are to be considered as a single atomic entity, none should be logically isolated from any of the other. They are always in co-active status during a life period called by APT (discussed later in this section) which is different for different element x of X (for a pre-fixed fuzzy decision maker).

For a fixed decision maker (intelligent agent), corresponding to every element x of X to belong to the fuzzy set A, there exists Atanassov Trio Functions i.e. a set of three AT functions. The value $\mu(x)$ in the fuzzy set A comes at the instant $t = T$ from the function $m(x, t)$. This function $m(x, t)$ finally converges at the value $\mu(x)$ after a course of sufficient growth. The function $m(x, t)$ gets feeding from $h(x, t)$ in a continuous manner starting from time $t = 0$ till time $t = T$. None else feeds $m(x, t)$. This idea will be more clear in the next subsection.

The membership value $\mu(x)$ for an element x in a fuzzy set A (or in an intuitionistic fuzzy set or in a soft computing set) can never be derived without the activation of AT functions in the cognition system, and this happens to any brain of human being or animal or of any living thing, irrespective of his education or knowledge.

3.2 Atanassov Initialization

At time $t = 0$ i.e. at the starting instant of time for evaluating the membership value $\mu(x)$, any decision process in the cognition system starts with AT functions with the following initial values:

$$\textbf{h}(\textbf{x}, \textbf{0}) = \textbf{1}, \text{with } \textbf{m}(\textbf{x}, \textbf{0}) = \textbf{0} \text{ and } \textbf{n}(\textbf{x}, \textbf{0}) = \textbf{0}.$$

The clock starts from time $t = 0$ and the whistle blows from this initialization only. This initialization $\langle 0, 0, 1 \rangle$ is called by 'Atanassov Initialization'.

It is important to understand that Atanassov Initialization is not initialized by any choice of the decision maker or by any decision of the decision maker or by any prior information from the kernel of the cognition system to the outer-sense of the decision maker. It is never initialized by the decision maker himself, but it gets automatically initialized at the kernel during the execution of any decision making process. By decision maker, we shall mean here a human or an animal or a bird or any living thing which has a brain.

Every decision making process for evaluating $\mu(x)$ can be viewed as a process solving an abstract type of mathematical model of an 'Initial Value Problem'.

3.3 *Impossible Types of Initialization*

A decision maker, be him highly intelligent or not, can never conclude the value of $\mu(x)$ at time $t = 0$ i.e. <u>at no cost of time</u>. He must need some amount of time $t > 0$, which could be infinitesimal small Δt or moderately small or a large amount. Imagine a case of any soft computing set model (for example: Fuzzy Set Theory) where the dimension $h(x, t)$ does not exist in the mathematical model of its notion proposed by Prof. Zadeh. In such a case too, it is quite obvious that **none** of the following three types of initializations can anytime happen (can be anytime possible) in the cognition system of the decision maker while evaluating the membership value $\mu(x)$:

Type-(i) $m(x, 0) = 0$ and $n(x, 0) = 1$ at time $t = 0$, but finally converging at $m(x, T) = \mu(x)$ and $n(x, T) = 1 - \mu(x)$ after time $t = T$ (>0).

Type-(ii) $m(x, 0) = 1$ and $n(x, 0) = 0$ at time $t = 0$, but finally converging at $m(x, T) = \mu(x)$ and $n(x, T) = 1 - \mu(x)$ after time $t = T$ (>0).

Type-(iii) $m(x, 0) = k$ and $n(x, 0) = 1 - k$, at time $t = 0$ where k is some initial scalar constant, and finally converging at $m(x, T) = \mu(x)$ and $n(x, T) = 1 - \mu(x)$ after time $t = T$ (>0).

It is because of the fact that whatever be the soft computing set theory under consideration of the decision maker (or no standard theory under consideration of the decision maker), the initialization is always the Atanassov Initialization only, can not be any alternative. Although $\mu(x)$ in fuzzy set theory does not have any scope of link with the hesitation part $h(x, t)$ which is also called by 'undecided part', but the software of the cognition system or brain has the in-built function $h(x, t)$ without which no decision process can initiate.

This was a fact in the stone age of the earth too, and will continue to remain as a fact for ever.

3.4 *Atanassov Trio Bags*

These are three imaginary bags. During the progress of decision making process with respect to the variable 'time' in the brain while evaluating the membership value $\mu(x)$, imagine that the values of AT functions are stored and updated continuously, with respect to time, in the three bags: h-bag, m-bag and n-bag, <u>always replacing their previous values</u>. These three bags are called by **Atanassov Trio Bags** (see Fig. 1).

It is obvious that at time $t = 0$ each of the Atanassov Trio Bags contains the value corresponding to Atanassov Initialization, not else. Immediately after that, the m-bag and n-bag start getting credited with zero or more amount of values continuously from the h-bag, subject to fulfillment of Atanassov Constraint at every

Fig. 1 Atanassov Trio Bags

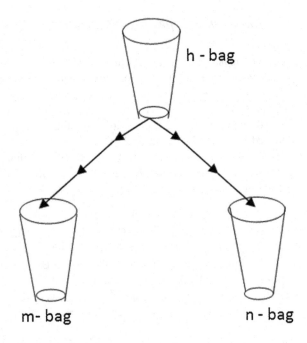

instant of time t assuming that the transaction time from h-bag to any bag is always nil. But there never happens a reverse flow, i.e. the h-bag does never get credited from any or both of m-bag and n-bag.

3.5 Atanassov Processing Time (APT)

While evaluating the membership value $\mu(x)$ of an element x, the Atanassov Initialization happens at time $t = 0$ at the human cognition system (or at the cognition system of the animal or bird whoever be the decision maker). At the very next instant of time, i.e. from time $t > 0$, the following <u>actions</u> happens simultaneously to the AT functions subject to fulfillment of Atanassov constraint (assuming that the transaction time from h-bag to any bag is always nil):

(i) $h(x, t)$ starts reducing (at least non-increasing), and
(ii) $m(x, t)$ as well as $n(x, t)$ start increasing (non-decreasing).

After certain amount of time, say after $t = T\ (>0)$ the processing of the decision making process stops (converges) at the following state, say:

$$h(x, T) = \pi(x), \text{ with } m(x, T) = \mu(x) \text{ and } n(x, T) = \vartheta(x)$$

where $\pi(x) + \mu(x) + \vartheta(x) = 1$, and after which there is no further updation happens to the values of AT functions in the cognition system. Then we say that for the complete evaluation of the membership value μ(x) by the concerned decision maker corresponding to the element x, the **Atanassov Processing Time (APT)** is T unit of time.

See at Fig. 2. The final values of the AT functions $\langle m(x,t), n(x,t), h(x,t)\rangle$ corresponding to the element x are given by:

$$m(x,T), n(x,T), h(x,T); \quad \text{where } APT(x) = T.$$

We call these final matured values $m(x,T)$ by $\mu(x)$, $n(x,T)$ by $\vartheta(x)$ and $h(x,T)$ by $\pi(x)$. They are respectively the **'membership value $\mu(x)$'**, the **'non-membership value $\vartheta(x)$'** and the **'hesitation value $\pi(x)$'** of the element x by the best possible judgement of the concerned decision maker in the Theory of IFS of Atanassov.

For a fixed decision maker, APT will be different in general for different elements of X. Also the equality $APT(x_1) = APT(x_2)$ does not necessarily mean that $\mu(x_1) = \mu(x_2)$ or that $\vartheta(x_1) = \vartheta(x_2)$ or that $\pi(x_1) = \pi(x_2)$, and conversely.

The following proposition can now be proposed.

Proposition 1

For evaluating the membership value μ(x) in a fuzzy set corresponding to any element x of the universe of discourse X, APT(x) > 0 irrespective of the quality or quantity of intelligence of the decision maker.

Justification

Whenever there is a evaluation process for μ(x) at the cognition system of the decision maker (human being or any living animal), it always starts with Atanassov Initialization

$$\text{"}h(x, 0) = 1 \text{ with } m(x, 0) = 0 \text{ and } n(x, 0) = 0\text{"} \text{ by default.}$$

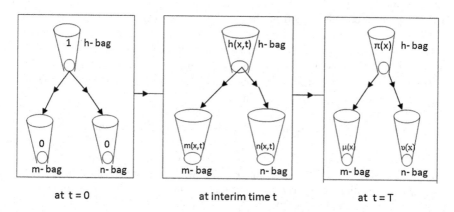

Fig. 2 Evaluation of μ(x) starting from 'Atanassov Initialization'

Atanassov Initialization happens at time t = 0, but thinking process of a decision maker for arriving at a decision about the value of μ(x) can not stop without any cost of processing time of the decision maker, whatever be the quality and quantity of excellence/intelligence of the decision maker. It may take an infinitesimal small positive time Δt or few ns or few micro seconds or minutes or hours or even more to arrive at a decision about the value of μ(x). In many cases it may stop at the same values "h(x, 0) = 1 with m(x, 0) = 0 and n(x, 0) = 0" of Atanassov Initialization i.e. at $\langle \mu(x) = 0, \vartheta(x) = 0, \pi(x) = 1 \rangle$, but after time T > 0 (not at time T = 0).

Example 1 Once in a tour to a jungle in Africa, Mr. Smith found five peculiar animals a_1, a_2, a_3, a_4, a_5 of probably different species, besides several known animals. Consider the universe of discourse $X = \{a_1, a_2, a_3, a_4, a_5\}$, and suppose that Mr. Smith is asked to name who among these five are of the species of 'cats'. For this, the decision maker (here Mr. Smith) attempts to construct an IFS A of X (where, at the end of the all necessary processing the IFS A could sometimes emerge to be a fuzzy set or even to be a crisp set, as the special cases of the IFS).

To construct the IFS A, the decision maker Mr. Smith considers elements of the universe X one by one, starting from the element a_1 (say).

Decision Process for the element a_1:

For evaluating the membership and non-membership values for the element a_1, suppose that the decision maker Mr. Smith starts his thinking process now (i.e. clock starts now for a_1) while t = 0. At this initial instant of time t = 0, the processing in the cognition system of the decision maker about the element a_1 initiates with Atanassov Initialization:

$$h(a_1, 0) = 1, \text{with } m(a_1, 0) = 0 \text{ and } n(a_1, 0) = 0.$$

Suppose that after $T_1 = 10$ min, the decision maker feels that his decision process is now over. His decision process is over because he could not get any clue by his best possible judgment about any further updation of the functions: membership function $m(a_1, t)$ and non-membership function $n(a_1, t)$, during the progress of decision process commencing from their initial values $m(a_1, 0) = 0$ and $n(a_1, 0) = 0$.

Consequently, APT(a_1) = 10 min and the final convergent values will stand at: $h(a_1, 10 \text{ min}) = 1$, with $m(a_1, 10 \text{ min}) = 0$ and $n(a_1, 10 \text{ min}) = 0$. i.e. according to Atanassov's IFS theory, $\mu(a_1) = 0, \vartheta(a_1) = 0$, and $\pi(a_1) = 1$ in the IFS A.

It may also happen that, although APT(a_1) = 10 min finally, but the following occurred with effect from time 9.2 min onwards:

$$h(a_1, 9.2 \text{ min}) = 1, \text{ with } m(a_1, 9.2 \text{ min}) = 0 \text{ and } n(a_1, 9.2 \text{ min}) = 0.$$

Until and unless the kernel of Mr. Smith is not convinced or satisfied by itself, the final convergence will not be guaranteed. An analogous situation happens in CPU while executing an algorithm, where the final results are already attained but it does not stop until and unless "STOP" be executed.

Decision Process for the element a_2:

For evaluating the membership and non-membership values for the element a_2, suppose that the decision maker Mr. Smith starts his thinking process now (i.e. clock starts now for a_2) while $t = 0$. At this initial instant of time $t = 0$, the processing in the cognition system of the decision maker about the element a_2 initiates with Atanassov Initialization:

$$h(a_2, 0) = 1, \text{ with } m(a_2, 0) = 0 \text{ and } n(a_2, 0) = 0.$$

Suppose that after $T_2 = 0.002$ min, the decision maker feels that his decision process is now over, and consequently the final convergent values stand at:

$$h(a_2, 0.002 \, \text{min}) = 0.7, \text{ with } m(a_2, 0.002 \, \text{min}) = 0 \text{ and } n(a_2, 0.002 \, \text{min}) = 0.3.$$

i.e. according to Atanassov's IFS theory, $\mu(a_2) = 0$, $\vartheta(a_2) = 0.3$, and $\pi(a_2) = 0.7$ in the IFS A, where $APT(a_2) = 0.002$ min.

Decision Process for the element a_3:

For evaluating the membership and non-membership values for the element a_3, suppose that the decision maker Mr. Smith starts his thinking process now (i.e. clock starts now for a_3) while $t = 0$. At this initial instant of time $t = 0$, the processing in the cognition system of the decision maker about the element a_3 initiates with Atanassov Initialization:

$$h(a_3, 0) = 1, \text{with } m(a_3, 0) = 0 \text{ and } n(a_3, 0) = 0.$$

Suppose that after $T_3 = 0.4$ min, the decision maker feels that his decision process is now over, and consequently the final convergent values stand at:

$$h(a_3, 0.4 \, \text{min}) = 0.8, \text{ with } m(a_3, 0.4 \, \text{min}) = 0.2 \text{ and } n(a_3, 0.4 \, \text{min}) = 0.$$

i.e. according to Atanassov's IFS theory, $\mu(a_2) = 0.2$, $\vartheta(a_2) = 0$, and $\pi(a_2) = 0.8$ in the IFS A, where $APT(a_3) = 0.4$ min.

Decision Process for the element a_4:

For evaluating the membership and non-membership values for the element a_4, suppose that the decision maker Mr. Smith starts his thinking process now (i.e. clock starts now for a_4) while $t = 0$. At this initial instant of time $t = 0$, the processing in the cognition system of the decision maker about the element a_4 initiates with Atanassov Initialization:

$$h(a_4, 0) = 1, \text{ with } m(a_4, 0) = 0 \text{ and } n(a_4, 0) = 0.$$

Suppose that after $T_4 = 3$ ns, the decision maker feels that his decision process is now over. But still he could not get any clue by his best possible judgment about any updation of the functions: membership function $m(a_4, t)$ and non-membership function $n(a_4, t)$, during the progress of decision process from their initial values $m(a_4, 0) = 0$ and $n(a_4, 0) = 0$.

Consequently, $APT(a_4) = 3$ ns and the final convergent values will stand at:

$$h(a_4, 3\,\text{ns}) = 1, \text{ with } m(a_4,\ 3\,\text{ns}) = 0 \text{ and } n(a_4, 3\,\text{ns}) = 0.$$

i.e. according to Atanassov's IFS theory, $\mu(a_4) = 0, \vartheta(a_4) = 0,$ and $\pi(a_4) = 1$ in the IFS A.

Decision Process for the element a_5:

For evaluating the membership and non-membership values for the element a_5, suppose that the decision maker Mr. Smith starts his thinking process now (i.e. clock starts now for a_5) while $t = 0$. At this initial instant of time $t = 0$, the processing in the cognition system of the decision maker about the element a_5 initiates with Atanassov Initialization:

$$h(a_5, 0) = 1, \text{ with } m(a_5, 0) = 0 \text{ and } n(a_5, 0) = 0.$$

Suppose that after $T_5 = 0.0005$ ns, the decision maker feels that his decision process is now over, and consequently the final convergent values stand at:

$$h(a_5, 0.0005\,\text{ns}) = 0.1, \text{ with } m(a_5, 0.0005\,\text{ns}) = 0.1 \text{ and } n(a_5, 0.0005\,\text{ns}) = 0.8.$$

i.e. according to Atanassov's IFS theory, $\mu(a_5) = 0.1, \vartheta(a_5) = 0.8,$ and $\pi(a_5) = 0.1$ in the IFS A, where $APT(a_5) = 0.0005$ ns.

The IFS A Proposed by Mr. Smith

Therefore, the IFS A proposed by Mr. Smith (decision maker here) will be as below by his best possible judgment:

$$A = \{(a_1, \langle 0, 0 \rangle), (a_2, \langle 0, 0.3 \rangle), (a_3, \langle 0.2, 0 \rangle), (a_4, \langle 0, 0 \rangle), (a_5, \langle 0.1, 0.8 \rangle)\}.$$

It signifies that according to the decision of Mr. Smith, a_5 may be 'Not Cat' and for other animals the matter is fully or 'almost fully' under hesitation (could not be decided finally).

The IFS A Proposed by Mr. Simmon

Consider another decision maker Mr. Simmon. Suppose that the IFS A proposed by Mr. Simmon is

$$A = \{(a_1, \langle 0, 0 \rangle), (a_2, \langle 0, 0.3 \rangle), (a_3, \langle 1, 0 \rangle), (a_4, \langle 0.2, 0 \rangle), (a_5, \langle 0, 1 \rangle)\}.$$

Then it signifies that according to the decision of Mr. Simmon, a_3 surely belongs to Cat species, a_5 surely does not belong to Cat species, and for other animals the matter is fully or 'almost fully' under hesitation (could not be decided finally).

The previous Proposition-1 is about a fuzzy decision maker. The following generalization is obvious for a decision maker using any soft computing set theory (Fuzzy Theory, L-Fuzzy theory, Type-2 fuzzy theory, intuitionistic fuzzy theory, Soft set theory, etc.).

Proposition 2

For evaluating the membership value $\mu(x)$ in any soft computing set model corresponding to any element x of the universe X, APT(x) > 0 irrespective of the quality and/or quantity of intelligence of the decision maker.

3.6 'Atanassov Speed of Convergence' (ASC) for an Element x

Suppose that there are 50 students in Class-V in Calcutta St. Paul's School. Let X be the set of all these 50 students, say $X = \{ x_1, x_2, x_3, \ldots, x_{50} \}$. Suppose that the decision maker Mr. Smith proposes an IFS A of all tall students of X, in which corresponding to the student x_1 the following is his best judgment:

$$m(x_1) = 0.80, \ n(x_1) = 0.05.$$

For converging to the above final scores, suppose that the time taken by the cognition processor of Mr. Smith is T_1. According to our hypothesis, at time t = 0 the Atanassov's Initialization is set by the Processor (Brain of Mr. Smith) as below:

$$h(x, 0) = 1, \ \text{with } m(x, 0) = 0 \text{ and } n(x, 0) = 0.$$

Immediately after Atanassov's Initialization, h(x, t) starts falling (at least non-increasing) and m(x, t), n(x, t) both start gaining (at least non-decreasing). The process continues for time T_1 after which there is no updation occur to the AT functions corresponding to this student x_1. Suppose that finally, at time t = T_1, the trio functions converge to the following scores:

$$h(x, T_1) = 0.15, \text{with } m(x, T_1) = 0.80 \text{ and } n(x, T_1) = 0.05.$$

Once the decision process stops at time t = T_1, the total finally 'decided part' out of initial hesitation amount (equal to 1) happens to be equal to:

$$m(x, T_1) + n(x, T_1) = 0.80 + 0.05 = 0.85,$$

while the rest amount 0.15 corresponds to the finally 'undecided part'.

The **Atanassov Speed of Convergence (ASC)** for an element x is a kind of **average intuitionistic speed** at which its trio functions $\langle m(x, t), n(x, t), h(x, t) \rangle$ converges to the final triplet $\langle m(x, T), n(x, T), h(x, T) \rangle$ i.e. the triplet $\langle \mu(x), \upsilon(x), \pi(x) \rangle$

where T is the value of APT(x), beginning from its initial triplet value $\langle 0, 0, 1 \rangle$ of Atanassov's Initialization, and is defined to be

$$ASC(x) = \frac{m(x, T) + n(x, T)}{T}, \quad \text{where T is the value of APT}(x).$$

One could view ASC as the "decision distance (membership and non-membership combined) covered in one unit of time".

The expression for ASC(x) consists of two mutually exclusive parts:

$$\frac{m(x, T)}{T} \quad \text{and} \quad \frac{n(x, T)}{T}.$$

The part $\frac{m(x, T)}{T}$ is called the **Atanassov Speed of m-Convergence (ASMC)**, and the part $\frac{n(x, T)}{T}$ is called the **Atanassov Speed of n-Convergence (ASNC)** for the element x.

$$\text{i.e.} \quad ASC(x) = ASMC(x) + ASNC(x).$$

Obviously, ASMC(x) and ASNC(x) are also kind of average intuitionistic speed. The trio speeds ASC(x), ASMC(x) and ASNC(x) are different, in general, for different elements of X and ofcourse different for different decision makers. For a fuzzy set or a crisp set, $ASC(x) = \frac{1}{T}$. However, for a crisp set, ASMC(x) = 1 and ASNC(x) = 0.

Note: Although APT(x) = T, it may be noted that the function m(x, t) may attain the value equal to m(x, T) sometimes at time $t = \tau$, which is little earlier than the Tth instant. But the decision making process does not stop at time $t = \tau$, and hence we can not regard the trio values $\langle m(x, \tau), n(x, \tau), h(x, \tau) \rangle$ to be the final convergent values, even though it is mathematically same as $\langle m(x, T), n(x, T), h(x, T) \rangle$.

Example 2 In the above example, if APT(x$_1$) is 10^{-1000} ns then

$$ASC(x_1) = 0.85/10^{-1000} \text{ unit/ns} = 85 \times 10^{998} \text{ unit/ns}.$$
$$ASMC(x) = 0.80/10^{-1000} \text{ unit/ns} = 80 \times 10^{998} \text{ unit/ns}.$$
$$\text{and} \quad ASNC(x) = 0.05/10^{-1000} \text{ unit/ns} = 5 \times 10^{998} \text{ unit/ns}.$$

3.7 Categorizing the 'Quality' of a Decision Maker

The quality of every decision maker is very important in the concerned organization. There is no absolutely precise mathematical model in the literature to measure the quality of a decision maker, probably because of the fact that it needs an establishment of a soft computing formula for Quality where the inputs are soft data besides crisp data. In the theory of CIFS we present a method to categorize the quality of a decision maker into three categories as: BAD, GOOD, and EXCELLENT.

The algorithm for categorizing the 'Quality' of a decision maker is mainly based upon three issues here:

(i)   How much amount is the decided part?
(ii) Out of the decided amount, how much is in favour of 'membership' part and how much is in favour of 'non-membership' part?
      (Obviously, it is nice to see that almost full amount of the decided amount happens to be either in favour of the 'membership' part or in favour of the 'non-membership' part, but not being distributed between these two).
(iii) How fast the decision is taken by the decision maker?

The number of adjectives (about the quality) here is taken to be three only: BAD, GOOD, and EXCELLENT; but can be extended up to any desired number.

For a **GOOD** decision maker (intelligent agent), the following two conditions are true for all or for almost all the elements x of X:

(i) $m(x, T) + n(x, T)$ will be 1 or very close to 1, and
(ii) $ASC(x)$ will be high for all or for almost all the elements of X.

For an **EXCELLENT** decision maker (intelligent agent), the following two conditions are true for all or for almost all the elements x of X:

(i) $\max \{m(x, T), n(x, T)\}$ will be 1 or very close to 1, and
(ii) $ASC(x)$ will be high for all or for almost all the elements of X.

For a **BAD** decision maker (intelligent agent), the following two conditions are true for all or for almost all the elements x of X:

(i) $\max \{m(x, T), n(x, T)\}$ will be 0 or very close to 0, whereas obviously $h(x, t)$ will be very close to 1, and
(ii) $ASC(x)$ will be low for all or for almost all the elements of X causing the processing slow.

3.8   Speed of Evaluation (Trio Speeds)

Let us assume that the decision maker (intelligent agent) is fixed. While proposing an IFS A of the universe X, corresponding to each element x each of the three AT functions starts getting updated with respect to time (which starts from t = 0) and finally converges (process of updation stops) after a finite quantity T (>0) of time.

Speed of Evaluation of the decision making process at variable time t exists for each of the AT functions, which are the functions $\frac{dm}{dt}, \frac{dn}{dt}$, and $\frac{dh}{dt}$.

These are also to be called the **trio speeds** corresponding to a given element x with respect to a given decision maker. They need not be equal in magnitude in general.

If APT(x) = T, then it is obvious that $\forall t \in [0, T]$, we have

$$\text{(i)}\ \frac{dm}{dt} \geq 0, \quad \text{(ii)}\ \frac{dn}{dt} \geq 0, \quad \text{and} \quad \text{(iii)}\ \frac{dh}{dt} \leq 0.$$

It may be noted that Atanassov Speed of Convergence is the <u>average</u> speed of complete processing for the evaluation of $\mu(x)$. It should not be confused with the trio speeds of the moving point $V(m(x, t), n(x, t), h(x, t))$ which starts from its initial point $P(0, 0, 1)$ and ends at the point of convergence $C(\mu(x), \upsilon(x), \pi(x))$.

For any decision maker (intelligent agent), the trio speeds are not consistent in general during the time-interval [0, T]. The path traversed by the variable point $V(m(x, t), n(x, t), h(x, t))$ is never a discontinuous curve, details of which is exercised in the subsequent subsections here. See Figs. 4, 5 and 6.

Proposition 3

For evaluating the membership and non-membership values of an element x, the rate at which h(x, t) decreases during the progress of the decision process with respect to time t in [0, T] where T is the APT(x), the m(x, t) and n(x, t) together grows exactly at the same rate.

Proof Atanassov constraint for a given element x at any variable time t is

$$m(x, t) + n(x, t) + h(x, t) = 1.$$

Differentiating w.r.t time t, we see that

$$\frac{d}{dt} m(x, t) + \frac{d}{dt} n(x, t) + \frac{d}{dt} h(x, t) = 0$$

$$\text{i.e.} \quad \frac{d}{dt} m(x, t) + \frac{d}{dt} n(x, t) = -\frac{d}{dt} h(x, t)$$

Both sides of this equality are non-negative values. This justifies the proposition.

However, the second derivatives $D^2 m(x, t), D^2 n(x, t)$, or $D^2 h(x, t)$ could be positive or negative or zero, but for any time t in [0, T] we have $D^2 m(x, t) + D^2 n(x, t) + D^2 h(x, t) = 0$. Thus at the end of the convergence process at Tth instant of time where T is the value of APT(x), the following results outcome:

$$\underset{t\to T}{Lt}\ m(x,t) = \mu(x),\quad \underset{t\to T}{Lt}\ n(x,t) = \vartheta(x)\quad \text{and}\quad \underset{t\to T}{Lt}\ h(x,t) = \pi(x),$$

such that

$$\mu(x) + \vartheta(x) + \pi(x) = 1.$$

The four variable parameters m, n, h and t could be viewed to form a 4-dimensional hyperspace in the Theory of CIFS. Since our interest is on the trio m, n and h only, for our next subsection here we consider 3-D geometry with three mutually perpendicular axes called by m-axis, n-axis and h-axis, forming a 3-D mnh-space.

3.9 Decision Trajectory Curve and Atanassov Point of Convergence

Any decision with respect to any real life task takes some amount of time. If something happens in zero amount of time, it can not be regarded as a decision process. In fact nothing in reality happens in zero amount of time. While evaluating $\mu(x)$, be it in a fuzzy set or in an IFS or in any soft computing set model, the decision does not complete in zero time, i.e. the decision does not yield a sudden value for $\mu(x)$ at time t = 0. For every such case, the starting point is at Atanassov Initialization trio $\langle 0, 0, 1 \rangle$ of the AT functions $\langle m(x,t), n(x,t), h(x,t) \rangle$

Consider a decision making process for evaluating $\mu(x)$. Let us imagine this initial trio $\langle 0, 0, 1 \rangle$ as a 3-D point P(0, 0, 1) in the mnh-space, and the trio at any time t as the 3-D variable point $V(m(x,t), n(x,t), h(x,t))$ in the mnh-space. Initiating from the point P(0, 0, 1) corresponding to time t = 0, the decision process progress in the cognition system of the decision maker, and accordingly the variable point $V(m(x,t), n(x,t), h(x,t))$ moves forward with time t. After time t = APT (x) = T, the variable point V stops at the point $C(m(x,T), n(x,T), h(x,T))$ i.e. at the point $C(\mu(x), \upsilon(x), \pi(x))$, because no further updation of the trio $\langle m(x,t), n(x,t), h(x,t) \rangle$ happens or will happen. The point V moves on a 3-D surface in mnh-space. It is justified later here that this surface in the mnh-space is nothing but a plane surface. This plane contains the triangle PQR with vertices P(0, 0, 1), Q(0, 1, 0) and R(1, 0, 0) which is called by **Atanassov Absolute Triangle**.

The corresponding curve-path PVC of the variable point V lying upon the plane of Atanassov Absolute Triangle is called the **Decision Trajectory Curve** for x. Surely it is always a curve of finite length, even if APT(x) be very large (Fig. 3).

The end point $C = C(\mu(x), \upsilon(x), \pi(x))$ is called the **Atanassov Point of Convergence** for x. Obviously, for different decision makers the Decision Trajectory Curve for the element x will be different in general (Figs. 4, 5 and 6).

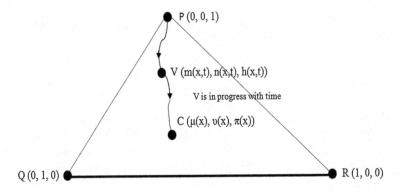

Fig. 3 Decision trajectory curve PVC of a decision making process for a intuitionistic fuzzy decision about $\mu(x)$

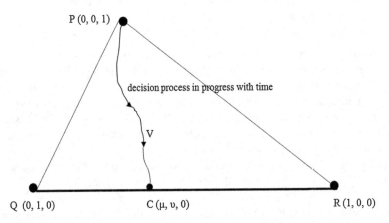

Fig. 4 Decision trajectory curve PVC of a decision making process for a fuzzy decision about $\mu(x)$

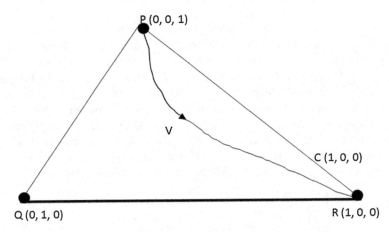

Fig. 5 Decision trajectory curve PVC of a decision making process for a crisp decision about $\mu(x)$

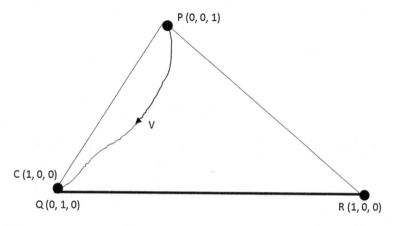

Fig. 6 Decision trajectory curve PVC of a decision making process for a crisp decision about μ(x)

3.10   Atanassov Absolute Triangle

During the course of evaluation of μ(x), the decision process progress initiating from the point P(0, 0, 1) at time t = 0, and accordingly the variable point $V(m(x, t), n(x, t), h(x, t))$ moves forward with time t. But the movement of the variable point V does always remain confined upon and within a triangular plane PQR in the mnh-space. The equilateral triangle PQR of side $\sqrt{2}$ unit in the mnh-space is called the **Atanassov Absolute Triangle** whose vertices are P(0, 0, 1), Q(0, 1, 0) and R (1, 0, 0).

Here the point P lies upon the h-axis, Q lies upon n-axis and R lies upon m-axis (see Fig. 7). The area of the Atanassov Absolute Triangle is $\sqrt{3}/2$ unit^2. It is obvious that whatever be the element x of X and whoever be the decision maker, Atanassov Absolute Triangle is a unique triangle, absolutely fixed for all cases, for all IFSs.

The side QR of length $\sqrt{2}$ unit is called the base of the Atanassov Absolute Triangle. Let us call this base by **Zadeh Line Segment**, in the name of the great philosopher Prof. Zadeh. For every fuzzy set, the point V will always terminate at some point lying upon the **Zadeh Line Segment**.

3.11   'Decision Line-Segments' and 'Atanassov Line-Segment' for an Element x

For any fuzzy or intuitionistic fuzzy problem, for a given element x the Decision Line-segments are the variable line-segments L_1L_2, each being of length $\sqrt{2}$ unit and parallel to the line segment QR. While evaluating the membership value μ(x), the

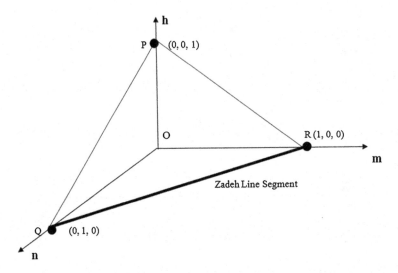

Fig. 7 Atanassov Absolute Triangle

process starts at time t = 0 with Atanassov initialization "h(x, 0) = 1 with m(x, 0) = 0 and n(x, 0) = 0" by default; and the **initial decision line-segment** is that line-segment which passes through the point P (0, 0, 1). The complete portion of the initial decision line-segment lies outside the Atanassov Absolute Triangle, except the point P which is common to both.

With the progress of time, the decision-line slides down from its initial position P, sometimes with slow pace and sometimes at faster pace; and finally at time t = APT(x) = T (>0) it stops at its position L_1ABL_2 (say) which passes through the Atanassov Point of Convergence $C(\mu(x), \upsilon(x), \pi(x))$. This last decision line-segment L_1ABL_2 is called the **Atanassov Line-Segment** for the element x. Although the Atanassov Line-Segment L_1ABL_2 is of length $\sqrt{2}$ unit, but the portion of it lying inside the Atanassov Absolute Triangle is AB of length $\sqrt{2} \cdot (1 - \pi(x))$ unit connecting the points $A(0, 1 - \pi(x), \pi(x))$ and $B(1 - \pi(x), 0, \pi(x))$. Obviously, for different decision makers the Atanassov Line-Segment for the element x will be different in general. However for every fuzzy set, the Atanassov Line-Segment will become coincident with the Zadeh Line Segment always, as a special case.

It may be calculated (using 3-D solid geometry in the variables m, n and h) that the equation of the 3-D line AB of Atanassov Line-Segment is:

$$m + n = 1 - \pi(x), \quad h = \pi(x).$$

In a fuzzy set, the Atanassov Line-Segment for every element x coincides with the Zadeh Line-Segment (as a special case) (Fig. 8).

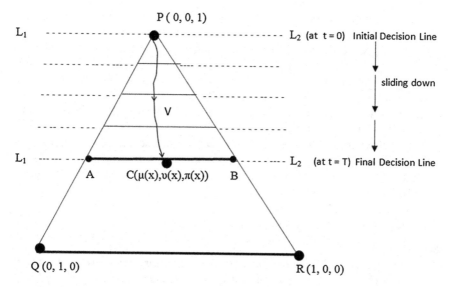

Fig. 8 Decision line-segment is sliding down with time, as the decision process is in progress starting from time t = 0

3.12 'Atanassov Triangle' for the Element x

The equilateral triangle PAB is called the **Atanassov Triangle** for the element x whose vertices are the points $P(0,0,1)$, $A(0, 1 - \pi(x), \pi(x))$ and $B(1 - \pi(x), 0, \pi(x))$.

Here the point $C(\mu(x), \upsilon(x), \pi(x))$ lies upon the line segment AB.

The side of this Atanassov Triangle is $\sqrt{2} \cdot (1 - \pi(x))$ unit and its area is $\sqrt{3}(1 - \pi(x))^2/2$ unit^2. For a fuzzy set, being a special case, Atanassov Triangle for x coincides with the Atanassov Absolute Triangle. Whatever be the element x or whoever be the decision maker, the Atanassov Absolute Triangle is a unique fixed triangle, whereas Atanassov Triangle is a variable triangle different for different elements and also for different decision makers.

Proposition 4
The decision trajectory curve PVC does never intersect the line-segment PC, except the possibility of tangential progress sometimes.

Justification:

At time t = 0 the decision process for the value of $\mu(x)$ initiates with Atanassov Initialization "h(x, 0) = 1 with m(x, 0) = 0 and n(x, 0) = 0" by default from the point P(0, 0, 1). There may be a tangential progress of curve PVC with the line segment

PC near the neighborhood of P initially, for some amount of time retaining the values of AT functions equal to the Atanassov Initialization values. However, after time T = APT(x) the decision process permanently stops at the Atanassov Point of Convergence $C(\mu(x), \upsilon(x), \pi(x))$.

Also, we know: max m(x, t) = $\mu(x)$, max n(x, t) = $\upsilon(x)$, and min h(x, t) = $\pi(x)$.

These extremum state occur at the point C (and also may be at many points on the curve only while there is a tangential progress of curve PVC with the line segment PC near the neighborhood of C).

If the decision trajectory curve PVC intersects the line-segment PC at least at one point $Z(m(x, \tau), n(x, \tau), h(x, \tau))$ at time t = τ where $\tau \in [0, T)$, then it is obvious that at least one of the three results: (i) max m(x, t) = $\mu(x)$, (ii) max n(x, t) = $\upsilon(x)$, and (iii) min h(x, t) = $\pi(x)$ can not be true.

Hence the proposition is justified.

The following proposition follows straightforward.

Proposition 5
The line-segment PC divides the Atanassov Triangle (for x) into two subtriangles PAC and PBC, one of which contains the decision trajectory curve PVC completely inside.

Proposition 6
For a fixed decision maker, if the Atanassov Line-segments for two distinct elements x_1 and x_2 of X coincide, then "APT(x_1) = APT(x_2)" $\Rightarrow$ ASC(x_1) = ASC(x_2). However, in general ASMC(x_1) $\neq$ ASMC(x_2) and ASNC(x_1) $\neq$ ASNC(x_2).

Proof Suppose that $C_1(\mu(x_1), \upsilon(x_1), \pi(x_1))$ is the point of convergence for x_1 and $C_2(\mu(x_2), \upsilon(x_2), \pi(x_2))$ is the point of convergence for x_2. Since the Atanassov Line-segments for them are coincident, they are the line segments AC_1B and AC_2B.

Clearly $\pi(x_1) = \pi(x_2)$, i.e. $\mu(x_1) + \upsilon(x_1) = \mu(x_2) + \upsilon(x_2)$; but in general $\mu(x_1) \neq \mu(x_2)$ and $\upsilon(x_1) \neq \upsilon(x_2)$.

Therefore, ASC(x_1) = ASC(x_2), but in general ASMC(x_1) $\neq$ ASMC(x_2) and ASNC(x_1) $\neq$ ASNC(x_2). Hence Proved.

3.13 *'General IF Case' and 'Five Special IF Cases' in Theory of CIFS*

Suppose that, on having complete processing of the membership value for the element x, the output result is $\mu(x)$; and the time consumed for this processing is APT(x) = T. Suppose that the Atanassov Absolute Triangle is the triangle PQR.

Then, the most general intuitionistic fuzzy case in CIFS for the element x is as below.

3.13.1 General IF Case

The Atanassov line-segment is the decision line-segment L_1ABL_2 passing through the points $A(0, 1 - \pi(x), \pi(x))$ and $B(1 - \pi(x), 0, \pi(x))$ at time $t = APT(x) = T(>0)$, where A lies the upon the side PQ and B lies upon the side PR of the Atanassov Absolute Triangle PQR (general Intuitionistic Fuzzy cases).

The following five are the special cases of intuitionistic fuzzy in the real life soft computing problems for an element x of X.

3.13.2 Five Special IF Cases

Case (1) *the Atanassov line-segment coincides with the 'initial decision line-segment' at time $t = T$.* (purely an Intuitionistic Fuzzy case).

Case (2) *the Atanassov line-segment coincides with the line-segment QR at time $t = T$.* (Special case of Intuitionistic Fuzzy: Fuzzy/Crisp).

The Case (2) has further special three subcases:

Subcase (2.1) *Atanassov Point of Convergence C coincides with the point Q.* (Intuitionistic Fuzzy case which is a **Crisp** case).

Subcase (2.2) *Atanassov Point of Convergence C coincides with the point P.* (Intuitionistic Fuzzy case which is a **Crisp** case).

Subcase (2.3) *Atanassov Point of Convergence C lies at an interim point on the line-segment QR.* (Intuitionistic Fuzzy cases which are the general form of all **Fuzzy** cases).

3.13.3 Annulus Sphere in the Cognition System of a Decision Maker

It is fact that the cognition system of a decision maker (fuzzy decision maker or intuitionistic fuzzy decision maker or crisp decision maker) can not evaluate the membership value $\mu(x)$ of an element x without initiating from the Atanassov Initialization "$h(x, 0) = 1$ with $m(x, 0) = 0$ and $n(x, 0) = 0$" by default, irrespective of his awareness/knowledge of IFS Theory (see Fig. 9).

It is because of the fact that this intuitionistic fuzzy processing happens at the kernel of the brain (CPU) of the decision maker, analogous to the case of execution of FORTRAN codes in CPU, irrespective of the awareness/knowledge of the concept of Machine Language by the concerned 'higher level language programmer' (see Fig. 10). Here the decision maker may be a fuzzy decision maker or any kind of decision maker (who may be a layman of IFS theory or of Fuzzy theory, or who could be even an animal or a living thing having brain).

Unfortunately, even for a highly intelligent fuzzy decision maker or for a highly intelligent intuitionistic fuzzy decision maker, it is impossible to know in advance whether by his best possible judgment the membership value $\mu(x)$ will finally come

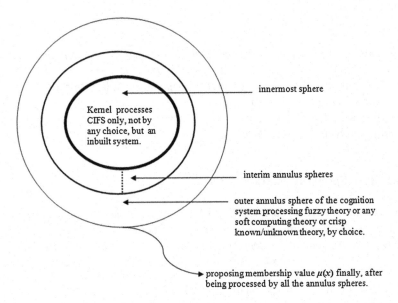

Fig. 9 Cognition System of a decision maker in CIFS, be it a human or animal or bird or any living thing which has brain

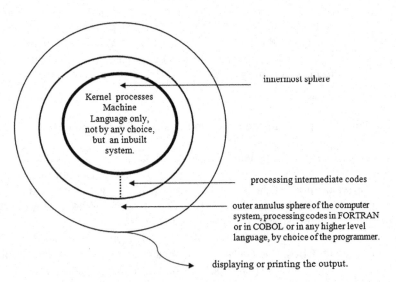

Fig. 10 CPU of a computer with common machine language at the kernel irrespective of any higher level language of the programmer by his own choice

out of the "General IF Case" or will come out of the special five other intuitionistic fuzzy cases, as mentioned above. Besides that, for different elements of X the cases (General case or Five Special case) are different as well as independent.

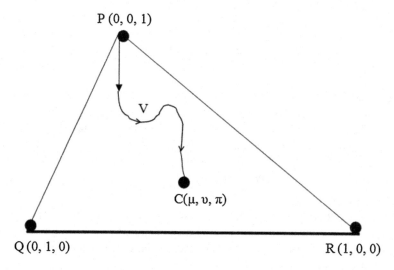

Fig. 11 An invalid decision trajectory curve

Consequently, the following curve PVC (see Fig. 11) is an invalid decision trajectory curve, which can never exist in any cognition system of any living thing.

Suppose that the orthogonal projection of the decision trajectory curve PVC on the mn-plane is the curve OV_0C_0, where V_0 the projection-point of the variable point $V(m(x, t), n(x, t), h(x, t))$ on the mn-plane. In between the extreme points O and C_0, the curve OV_0C_0 may run sometimes parallel to m-axis and sometimes parallel to n-axis too, but subject to the condition stated in earlier Propositions.

The following proposition is straightforward now.

Proposition 7

The projection of the decision trajectory curve PVC on the mn-plane is a curve OV_0C_0 which may have points of inflexion but its extreme points O and C_0 are respectively the point of minimum and the point of maximum. However, the curve OV_0C_0 may have infinite number of points of minimum and points of maximum, only at the neighborhood of these extreme points O and C_0 respectively.

The next proposition unearths the hidden truth

(i) about the Atanassov Constraint

$$h(x, t) + m(x, t) + n(x, t) = 1 \text{ for any time } t, \text{ and}$$

(ii) about the Atanassov Initialization

$$h(x, 0) = 1, \text{ with } m(x, 0) = 0 \text{ and } n(x, 0) = 0, \text{ and also}$$

(iii) about the intuitionistic fuzzy degrees $\langle \mu(x), \vartheta(x), \pi(x) \rangle$ of x,
corresponding to every decision maker using any soft-computing set theory of his personal choice like: Fuzzy set theory, Intuitionistic fuzzy set theory

(vague sets are nothing but intuitionistic fuzzy sets, as justified and reported by many authors), i–v fuzzy set theory, L-fuzzy set theory, Rough set theory, Soft set theory, etc. or even using crisp logic or any other logic (unknown to us presently).

Proposition 8

For any decision maker, be it a human or an animal or any living thing which has brain, it is impossible that his brain (kernel of his cognitive system) does always have the indeterministic component (i.e. the hesitation component or undecided component) h(x, t) to be nil for the element x of the universe X, while going to propose the corresponding membership value μ(x).

Proof Suppose that APT(x) = T.

Consider the 2-D curve m = m(x, t) on tm-plane (as shown in Fig. 12). Suppose that A_m is the area under the curve m = m(x, t) bounded by the lines t-axis, m-axis and the line t = T. Clearly, corresponding to a given decision maker A_m is not a function of t but of the parameter x.

Now consider the 2-D curve n = n(x, t) on tn-plane (as shown in Fig. 13). Suppose that A_n is the area under the curve n = n(x, t) bounded by the lines t-axis, n-axis and the line t = T.

Also consider the 2-D curve h = h(x, t) on th-plane (as shown in Fig. 14).

Suppose that A_h is the area under the curve h = h(x, t) bounded by the lines t-axis, h-axis and the line t = T.

Now consider the Atanassov Constraint

$$m(x, t) + n(x, t) + h(x, t) = 1.$$

Integrating with respect to time t we get,

$$\int_0^T m(x, t)dt + \int_0^T n(x, t)dt + \int_0^T h(x, t)dt = T$$

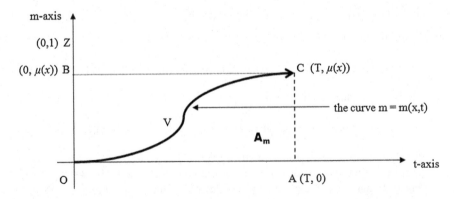

Fig. 12 The curve m = m(x, t) on tm-plane

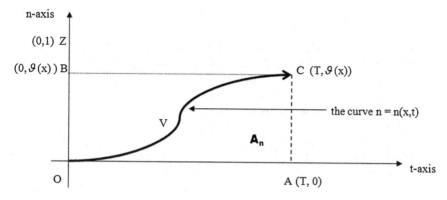

Fig. 13 The curve n = n(x, t) on tn-plane

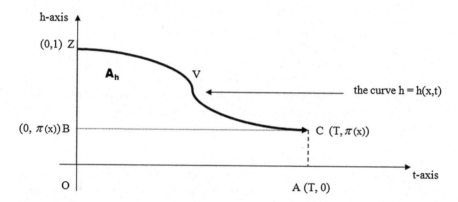

Fig. 14 The curve h = h(x, t) on th-plane

or,

$$A_m + A_n + A_h = T \qquad (1)$$

Now let us agree that in this real world it is quite obvious that nothing can happen or can be achieved at the cost of zero amount of time. Even for light particle to travel one nano centimeter, it takes some infinitesimal small amount of time Δt which is greater than zero. Consequently, in Fig. 12 it is obvious that the curve OVC does always starts from the origin, for which t = 0 and m = 0.

Thus, from the Fig. 12, it is obvious that

area amount $A_m <$ Area of the rectangle OACB.

(There is no chance under any circumstances that the Area amount $A_m =$ Area of the rectangle OACB, unless both are equal to zero).

Therefore, $A_m < T \cdot \mu(x)$ in general, excluding the case $\mu(x) = 0$ for which the equality $A_m = T \cdot \mu(x)$ holds good.

In a similar way it can be established that $A_n < T \cdot \vartheta(x)$ in general, excluding the case $\vartheta(x) = 0$ for which the equality $A_n = T \cdot \vartheta(x)$ holds good.

And also it is true that $A_h > T \cdot \pi(x)$ in general, excluding the case $\pi(x) = 1$ for which the equality $A_h = T \cdot \pi(x)$ holds good.

Consider now the following three cases:

Case(1) $\mu(x) = 0$ and $\vartheta(x) > 0$. and
Case(2) $\mu(x) > 0$ and $\vartheta(x) = 0$.
Case(3) $\mu(x) > 0$ and $\vartheta(x) > 0$.

It is obvious that for all of these three cases

$$A_m + A_n < (T \cdot \mu(x) + T \cdot \vartheta(x)) \tag{2}$$

Now let us prove our proposition **by contradiction**.

For this, let us suppose that:

For any decision maker, be it a human or an animal or any living thing, it is **possible** *that his brain (kernel of his cognitive system) does always have the indeterministic component (i.e. the hesitation component or undecided component) h(x, t) to be nil while going to propose the membership value μ(x).*

Therefore, $h(x, t) = 0$ for every $t \in [0, T]$.

From (1), $T = A_m + A_n$.

Therefore, $T < (T \cdot \mu(x) + T \cdot \vartheta(x))$, using (2).

This means that $\mu(x) + \vartheta(x) > 1$, which is not possible in any soft-computing set theory (for instance, not possible in Fuzzy Set theory). Hence the Proposition. However for $t = 0$, the Atanassov Constraint $m(x, 0) + n(x, 0) + 0 = 1$ itself contradicts, and thus justifies the proposition directly.

4 Is 'Fuzzy Theory' an Appropriate Tool for Large Size Problems?

It is observed that $0 \le \pi(x) \le 1$ for every x of the universe X, whoever be the decision maker. As a special case, it may happen for one or few elements in the IFS A that $\pi(x) = 0$. But in ground reality, for a decision maker by the best possible processing in his cognition system, the data '$\pi(x) = 0$' can not be true in general for all and across all the elements x of any universe X while proposing an IFS A of X (as justified further in **Proposition 8** above). Even if it be true for one or few or many elements, it is illogical to believe that it is true for all and across all the elements of any universe X while proposing an IFS A.

Further to that, any real life soft-computing problem on this earth usually occurs involving more than one universe. There could be r number of universes viz. $X_1, X_2, \ldots, X_r$ in a given problem under consideration by a decision maker. And in that case it is extremely illogical to believe that '$\pi(x) = 0$' is true for all the elements of all the r universes. Neither any real logical system(s) nor the Nature can force a decision maker (human being or animal or any living object having a brain) either to stand strictly at the decision: "$\pi(x) = 0$ for every x of every X", or "to abandon his decision process otherwise".

Any decision process for deciding the membership value $\mu(x)$ starts with Atanassov's initialization $\langle 0, 0, 1 \rangle$ and then after certain amount of time T (called by Atanassov Processing Time) it converges to the trio $\langle \mu(x), \vartheta(x), \pi(x) \rangle$ without any further updation of the AT functions. In general in most of the cases, $\pi(x) \neq$ NIL. Even if $\pi(x) =$ NIL for one or few elements, it is a particular situation that $\pi(x)$ will be Nil for all the elements x of X in the IFS proposed by the decision maker. It is a very very special case, the following justification will support further to this hypothesis:

Suppose that, to solve an ill-defined problem we have to consider 10 number of universes, while in each universe there is at least 100 elements. Suppose that, to solve this problem an intelligent decision maker (say, a soft computing human expert) needs to consider more than 20 fuzzy sets of each universe. Clearly, he has to propose membership values by his best possible judgment for more than 20,000 elements. For deciding the membership value for each of these 20,000 elements the cognition system of the decision maker by default begins with Atanassov's initialization $\langle 0, 0, 1 \rangle$ and after certain time converges to the decision about $\mu(x)$ for the element. Can we presume that for each of these 20,000 elements his convergence process starting from the Atanassov's initialization trio $\langle 0, 0, 1 \rangle$ will stop at the trio $\langle \mu(x), \vartheta(x), 0 \rangle$ with $\pi(x) = 0$ for each and every x? Can we presume that there is not a single element x out of 20,000 elements for which the convergence process ends with some non-zero amount of $\pi(x)$?

This surely justifies that it may not be appropriate to use fuzzy theory if the problem under study involves the estimation of membership values for <u>large</u> number of elements of one or more universes.

In our everyday life, every human being is a decision maker at every moment of time (ignoring his sleeping period at night) and he is compelled to decide every day for large number of imprecise problems of various nature. But one can not be an excellent outstanding decision maker for all the unknown (or known) and unpredictable heterogeneous problems he faces every day without any hesitation at all. And this is true for all the decision makers on this earth.

Consequently, it is now a well justified fact that for a large or moderate size soft computing problem, it may not be appropriate to use the tool 'Fuzzy Theory' in order to get excellent results.

However, there are also a number of real life cases where only the best/excellent decision makers (in many cases pre-selected or pre-chosen) are allowed to take decisions who can do the job and are supposed to do the job **without any hesitation** on any issue of the problem under consideration, i.e. '$\pi(x) = 0$' is to be true everywhere during the execution of the problem-solving.

Consequently, at the end of this article, two examples of decision making problems (with methods of finding their solutions) are presented, out of which one example (see Sect. 6) will show the dominance of the application potential of 'intuitionistic fuzzy set theory' over 'fuzzy set theory', and the other example will show the converse i.e. the dominance of the application potential of 'fuzzy set theory' over 'intuitionistic fuzzy set theory' in some cases (for example, see Sect. 7).

5 Ordering (or Ranking) of Elements in an IFS on the Basis of Their Amount of Belongingness

There are several methods proposed by various authors on ranking of n number of IFSs of a universe, on ranking of n number of intuitionistic fuzzy numbers, etc. But, there is no method available in the existing literature for ordering (or, ranking) the elements of a given IFS A of the universe of discourse X on the basis of their amount of belongingness. It is to be noted that if A and B are two IFSs of a common universe of discourse X, then in general ranking of the elements in the IFS A will not be same as the ranking of the elements in the IFS B. It is because of the reason that ranking is to be done with respect to the property (properties) of the IFS, mainly with respect to the depth of belongingness of the elements into the concerned IFS.

5.1 An Example Showing Application Potential of 'Ordering (or Ranking) of Elements'

Consider the set X of all 50 students of Class-V of Calcutta High School. For an important school event, the Principal of the school seeks the list of all the students of Class-V who are smart. Suppose that an intelligent teacher Mr. Z proposes by his best possible judgment an IFS A of X to the Principal. Having this IFS A, the Principal is not feeling comfortable as he is a man of the subject 'Music'. He wants more precise information, i.e. the name of the most smart student, second most smart student, third most smart, … and so on. Now the problem is: How to rank/order the students in the merit of 'smartness'? What is the mathematical formula or model which can yield this result?

For this, a new powerful indicator called by **Atanassov Indicator** is introduced, applying which we can order the elements of the IFS A, according to the property of A and mainly with respect to the depth of belongingness of the elements into the IFS A.

5.1.1 Intuitionistic μ-index (iμi)

Consider an element x in the IFS A having the intuitionistic fuzzy degree $(x, \langle \mu(x), \vartheta(x), \pi(x) \rangle)$. The ratio of the 'membership value amount' to the 'total decided amount' (i.e. excluding the hesitation amount) is called the **intuitionistic μ-index (iμi)** of x denoted by iμi(x).

It signifies the percentage amount of $\mu(x)$ value in the total decided amount $(\mu(x) + \upsilon(x))$.

Thus, $i\mu i(x) = \frac{\mu(x)}{\mu(x) + v(x)}$ or, may be signified as $\frac{\mu(x)}{\mu(x) + v(x)} \times 100$ %.

In a fuzzy set, iμi(x) is equal to $\mu(x)$. In a crisp set iμi(x) is equal to 1.

Because of the non-zero hesitation part (as a special case it could be zero for few or all the elements) in the intuitionistic fuzzy degree $(x, \langle \mu(x), \vartheta(x), \pi(x) \rangle)$ of any element x in the IFS A, it is not a straightforward task to compare two given elements x_1 and x_2 to conclude which one of these two is more penetrated inside the region of IFS A i.e. which one of these two elements x_1 and x_2 carry more amount of depth of belongingness in the IFS A, satisfying the property of the object A. There must exist an intuitionistic fuzzy mathematical instrument which takes x_1 and x_2 as the inputs and yields the answer. In fact the instrument/indicator must be able to choose that element which has high amount of $\mu(x)$ value as well as high amount of iμi(x) value. The indicator AI has the potential to do this job appropriately.

5.1.2 Atanassov Indicator (AI)

For an element x in the IFS A of the universe X, the **Atanassov Indicator (AI)** of x is defined by:

$$AI(x) = \mu(x) \times i\mu i(x).$$

If, for two given elements x_1 and x_2 we have $AI(x_1) > AI(x_2)$, then we conclude that x_1 carries 'more amount of depth of belongingness' than x_2 to belong to the IFS A. However, if $AI(x_1) = AI(x_2)$ then there is a tie at this instant of time. It may be observed that for fuzzy sets, $AI(x) = [\mu(x)]^2$; and for crisp sets $AI(x) = 1$. With no loss of generality, we can choose $AI(x) = \mu(x)$ for the case of fuzzy sets, instead of $[\mu(x)]^2$. We claim that AI will play a major role in the subject of "Decision Sciences" in due time. We shall apply this indicator in our subsequent section in solving a problem of object recognition, one of the most important decision making problems on this earth.

Example 3 Suppose that there are 50 students in Class-V in Calcutta St. Paul's School. Let X be the set of all these 50 students, say $X = \{x_1, x_2, x_3, \ldots, x_{50}\}$. Consider the three students out of these 50 students named by: Jim, Tina and John. Suppose that in the IFS A of all smart students of Class-V in Calcutta St. Paul's School proposed by the teacher Z, the intuitionistic fuzzy degrees of these three students are as below:

$(Jim, \langle 0.6, 0.3, 0.1 \rangle), (Tina, \langle 0.5, 0.1, 0.4 \rangle)$ and $(John, \langle 0.7, 0.3, 0 \rangle)$.

It may be now calculated that AI(Jim) = 0.40, AI(Tina) = 0.4166 and AI (John) = 0.49 in the IFS A. Thus John is most smart boy among these three students, Tina is the second and Jim is third.

It may be noted that the result µ(Tina) < µ(Jim) does not necessarily always mean that Tina is less smart than Jim in the Theory of IFS.

However, in Fuzzy Theory as the special case of IFS Theory, the result µ(Tina) < µ(Jim) signifies that 'Tina is less smart than Jim'.

In the next sections we present two examples of 'decision making problems' with solutions out of which one example (in Sect. 6) will show the dominance of the application potential of intuitionistic fuzzy set theory over fuzzy set theory, and the other (in Sect. 7) will show the converse i.e. the dominance of the application potential of fuzzy set theory over intuitionistic fuzzy set theory (where decision makers are pre-selected and most intellectual in the context of the subject pertaining to the concerned problem).

6 An Application Domain to Understand the Potential of Intuitionistic Fuzzy Theory Over Fuzzy Theory

A huge amount of application scopes of fuzzy theory have been reported in literatures during the last four decades. The Intuitionistic Fuzzy Theory too has been applied in various problems in almost all areas of Science, Technology, Social Science, Law, Medical Science, etc. to list a few only out of many. In this section one example on application of intuitionistic fuzzy theory is presented for the sake of understanding the potential of intuitionistic fuzzy theory over fuzzy theory.

6.1 Object Recognition: An Impossible Task Without CIFS

In this section we propose an intuitionistic fuzzy hypothesis which we claim to be the 'decoding of the real steps executed in the brain while recognizing an unknown object'. However, the decoded algorithm could be an Open Topic of IFS Theory for rigorous further investigation by the researchers/scientists and experts, in particular of the areas of Biomedical Engineering, Brain, Neuroscience, etc.

We propose the hypothesis here by an example only. This example shows a huge role of Atanassov Indicator (AI) in decision making problems.

Consider the case of object recognition system by human beings with visual inputs by eyes. When you see a table by your eyes, you immediately (in time t > 0) recognize it and say that it is 'table' and/or you also may say that it is a furniture, etc. rejecting all other infinite number of possibilities like: it is a dog, book, banana, car, water drop, moon, an integer number 54, a house, chair, laptop, etc. When you see a banana by your eyes, you immediately (in time t > 0) recognize it and say that it is 'banana', rejecting all other infinite number of possibilities like: it is a dog, book, table, car, water drop, moon, an integer number 54, a house, chair, laptop, etc. It can be easily proved that the total number of objects you can recognize is infinite (as a proof, consider the fact that you can recognize any given integer number. Hence proved.).

Suppose that in and around us we have the following N number of sets of various known objects (not necessarily mutually disjoint) where N could be finite or countable infinite, and cardinality of each set could be finite or countable infinite or uncountable infinite.

(1) S_1 = set of all birds
(2) S_2 = set of all fishes
(3) S_3 = set of all furniture
(4) S_4 = set of all colors
(5) S_5 = set of all animals
(6) S_6 = set of all male people
(7) S_7 = set of all numbers
(8) S_8 = set of all books
(9) S_9 = set of all snakes
(10) S_{10} = set of all cups
(11) S_{11} = set of all sports
...
...
...
(N) S_N = set of all vehicles

Call the class of these sets as the **Universal Class C**.

For the sake of explanation here, consider a decision maker who is a human being named by Mr. D. By his best possible judgment, Mr. D can recognize any element of any of the above N sets but up to certain extent; for completely unknown object it is 0 %, for a completely sure case it is 100 %, and for any other case with some doubt, it is in the range of open interval (0 %, 100 %).

Our main interest here is NOT JUST to understand: "How does Mr. D recognize an object?" Our main interest here is to understand mainly: "What steps happen at the kernel of the cognition system of D so that D finally gives his decision on the recognition of an unknown object with visual inputs through his eyes?"

Pattern Recognition or Object Recognition is one of the earliest and important problems to the earth. We need to recognize new diseases, new stars, new planets, new solar systems, new viruses, new bacteria, etc. to list here a few only out of infinity for the human welfare, ultimately for the welfare of our planet earth. Recently, scientists have recognized the existence of boson but still they are in the process of recognizing whether it is all Higgs bosons or there are other bosons too. There are different theoretical methods every year being proposed and also being experimented by the researchers for well recognition of unknown objects. But whatever be the methods or whatever be the thoughts, we propose the hypothesis that all these are at the higher level of kernel of human cognition system, whereas at the core kernel it is always CIFS (analogous to the fact that: there could be several higher level language like FORTRAN, COBOL, BASIC etc., but the execution at CPU is always of machine codes only).

Suppose that D sees an unknown object O which he needs to recognize. i.e. D has to decide what is this object O. It is obvious that finally, after some time $T(>0)$, the object may appear to be recognized successfully by D as an object known to D according to the knowledge of D or may even remain fully unknown (having 100 % doubt) or may appear to be partially known (some amount of doubt).

We propose the following hypothesis as the solution-steps actually followed in the cognition system of D.

There are three main phases, which are executed sequentially in the cognition system of D.

Phase-1 **Shortlisting the Sets**
Shortlisting the i number of Sets $\dot{S}_1, \dot{S}_2, \dot{S}_3, \ldots, \dot{S}_i$ out of N sets of the Universal Class $\acute{C}$, where $0 \le i \le N$.

Phase-2 **Shortlisting the Elements**
Shortlisting the elements $o_{i1}, o_{i2}, o_{i3}, \ldots$ out of all the elements of the set $\dot{S}_i$, for every i.

Phase-3 **Result about Recognition** (Yes, No, Possibly, etc. cases)

CS Action in Phase-1 (Shortlisting the Sets)

The cognition system (CS) of the decision maker D first of all thinks about the set or sets of the Universal Class $\acute{C}$ to which the unknown object O belong. For this the CS wants to take an intuitionistic fuzzy set A of the Universal Class $\acute{C}$. And, for this the CS needs to estimate the trio values $\langle \mu(S_r), \vartheta(S_r), \pi(S_r) \rangle \; \forall r = 1, 2, 3, \ldots, N$.

For each of these estimation, the processing at the cognition system starts with Atanassov Initialization $\langle 0, 0, 1 \rangle$, and finally converges at $(S_r, \langle \mu(S_r), \vartheta(S_r), \pi(S_r) \rangle)$ where $r = 1, 2, 3, \ldots, N$.

Now consider the following three cases of Phase-1:

Case-1.1:

In the IFS A, there will be many sets S_r having trio values like $(S_r, \langle 0, 1, 0 \rangle)$ or $(S_r, \langle 0, 0, 1 \rangle)$ for many values of r. All these sets are to be ignored, i.e. not to be included in the shortlist. If it happens for all the N sets, then the recognition process stops without further processing with the result 'fully unknown'. Otherwise go to Case-1.2.

Case-1.2:

There is a possibility of nil or one or few (but not many) sets having trio values like $(S_r, \langle 1, 0, 0 \rangle)$ for some values of r. All these sets S_r are to be included in the shortlist. It is sure that this situation can not happen for all of the N sets of the universal class C. If there is at least one set S_r, then Go to Phase-2 now, ignoring Case-1.3.

Case-1.3:

If there is no such sets with $(S_r, \langle 1, 0, 0 \rangle)$ in Case-1.2, consider collection of those sets for which $AI(S_r) \geq \alpha$, where $\alpha \in [0, 1]$ is so chosen that this collection is not big enough, but at the same time sufficient thinking is to be done so that it is <u>not null-set too</u>.

In this way, the CS has completed the shortlisting of i number of Sets $\dot{S}_1, \dot{S}_2, \dot{S}_3, \ldots, \dot{S}_i$ out of N sets of the Universal Class C, where $0 \leq i \leq N$.

Suppose that C_S denotes the class of shortlisted sets $\dot{S}_1, \dot{S}_2, \dot{S}_3, \ldots, \dot{S}_i$. Clearly C_S is a subclass of the universal class C, with special cases that it could be equal to the universal class C itself or it could be the null set.

CS Action in Phase-2 (Shortlisting the Elements)

This phase commence with the class C_S as the main input, where it is not a null set. The cognition system (CS) of the decision maker D considers the shortlisted sets $\dot{S}_1, \dot{S}_2, \dot{S}_3, \ldots, \dot{S}_i$, one by one.

FOR j = 1, 2, 3,…, i, DO the following of Phase-2:

(i) Consider the set $\dot{S}_j = \{o_{j1}, o_{j2}, o_{j3}, \ldots\}$. The CS first of all thinks about the element or elements of the set $\dot{S}_j$ to which the unknown object O does match. According to the relevant properties, the CS wants to takes an intuitionistic fuzzy set A_j of the set $\dot{S}_j$. For this the CS needs to estimate the trio values $\langle \mu(o_{jk}), \vartheta(o_{jk}), \pi(o_{jk}) \rangle$ $\forall k$.

For each of these estimation, the processing at the cognition system starts with Atanassov Initialization $\langle 0, 0, 1 \rangle$, and finally converges at $(o_{jk}, \langle \mu(o_{jk}), \vartheta(o_{jk}), \pi(o_{jk}) \rangle)$ $\forall k$ by the best possible judgment of D (by his CS).

(ii) Now consider the following three cases:

Case-2.1:

In the IFS A_j, there will be many elements having trio values like $(o_{jk}, \langle 0, 1, 0 \rangle)$ or $(o_{jk}, \langle 0, 0, 1 \rangle)$ for many values of k. All these elements are to be ignored, not to be included in the shortlist. If it happens for all the values of k, then the recognition process stops without further processing with the result 'O is None from $\dot{S}_j$'. Otherwise go to Case-2.2.

Case-2.2:

There is a possibility of nil or one or few (but not many) elements having trio values like $(o_{jk}, \langle 1, 0, 0 \rangle)$ for some values of k. All these elements o_{jk} are to be included in the shortlist. It is sure that it can not happen for all of the elements of this set. If there is at least one element, then the recognition process stops without further processing with the result 'O is successfully recognized to be o_{jk}'. Otherwise go to Case-2.3.

Case-2.3:

If there is no such elements with $\left(o_{jk}, \langle 1, 0, 0 \rangle\right)$ in Case-2.2, consider collection of those elements for which $AI\left(o_{jk}\right) \geq \beta_j$, where $\beta_j \in [0, 1]$ is so chosen that this collection is not big enough, but at the same time sufficient thinking is to be done so that it is <u>not null-set too</u>.

(iii) In this way, the CS has completed the shortlisting of q number of elements $o_{j1}, o_{j2}, o_{j3}, \ldots o_{jq}$ out of all the elements of the set $\dot{S}_j$.

CS Action in Phase-3 (Result about Recognition)

  (i)   For every shortlisted element õ, Calculate AI(õ).
 (ii)   Now arrange all AIs in descending order of magnitude.
(iii)   If the first AI is equal to 1, it signifies that the object has been recognized successfully.
 (iv)   If the first AI is equal to 0, it signifies that the object can not be recognized.
  (v)   If the first AI is equal to α where $0 < \alpha < 1$, it signifies that the object is not recognized successfully, but partially. If the first p number of elements have equal or almost equal AIs, then any of these p elements have the possibility.

6.2 Explaining the Method by an Instance

Let us explain the method by an instance here. Suppose that Mr. D has seen a Dog which he has to recognize. Although it is a very trivial and simple case to recognize, nevertheless the brain of Mr. D takes some amount of time (whatever small it be) to complete the job.

There are three main phases, which are executed sequentially in the cognition system of Mr. D.

Phase-1: **Shortlisting the Sets**.
Phase-2: **Shortlisting the Elements**.
Phase-3: **Result about Recognition** (Yes, No, Possibly, etc. cases).

The following sequence of events happens in reality:
In Phase-1, the shortlisted class C_S of sets will at least contain the set S_5.
In Phase-2, out of all the elements of S_5 there will be an element in the IFS like $(DOG, \langle 1, 0, 0 \rangle)$.
And finally in Phase-3, Mr. D can give his decision that the unknown object is recognized to be a "dog".

7   An Example of Application Domain to Understand the Potential of Fuzzy Theory Over Intuitionistic Fuzzy Theory in Some Cases

Since its inception in 1965, Fuzzy Theory has been fluently applied in various problems in almost all areas of Science, Technology, Social Science, Law, Medical Science, etc. to list a few only out of many. But most probably no work has been so far reported in literature showing the application of fuzzy theory in the domain of Sports, in particular in most popular sports like Football, Cricket, Tennis, Badminton, Chess, etc. Statistical data and statistical analysis are commonly practiced in almost all areas of sports. Football is probably the most popular sports in the world, and every football game during every minute of play generates continuously the real time data for statistical analysis and conclusions. For the sake of presenting here one example on application of fuzzy theory, we choose Sports domain and here it is Football. In this sports area, the Referees are not a simple decision makers but supposed to be truly experts, the best available decision makers in this area selected and recruited by the Football Associations or clubs.

Consequently, we have to presume the following:

(i)   there is **no hesitation part** corresponding to any decision taken by the Referees,
(ii)   Referees' decisions are final.

Such type of problems where decision makers are most intellectual and knowledgeable on the concerned subject, where decision makers are pre-selected by top experts as genuine, excellent and extra-ordinary decision makers, it is surely the 'Fuzzy Theory' to be regarded as the most appropriate tool to deal with, even compared to the 'Intuitionistic Fuzzy Theory'.

In this application domain we consider only those cases of football matches for which the penalty shootout is the existing method for determining the Winner, otherwise that would have been drawn or tied after 90 min of play. The existing method of penalty shootout is having a lot of criticisms due to its huge demerits. In this research work the author proposes an improved method to FIFA (IFAB) to select the winner by a continuous soft-computing evaluation method. The method is called by **CESFM** which is basically a Fuzzy constructive method. But CESFM in its theory does not exclude penalty kicks by each team as followed presently after 90 min of play. The result of penalty kicks is one of the many components in the method of CESFM. This proposed fuzzy method can easily extract the true dominant team out of the two tied teams using a software called by CESFM-software with the help of continuous real time inputs from the Referee where some of the inputs are given by the Referee using his '**fuzzy pocket machine**' M which is an electronic wireless machine (looking like a mobile phone). It is claimed that if FIFA (IFAB) implements this method of **CESFM** in World Cup Football matches, it will be a true justice to the football world, to the fans, to the players, to the teams (in particular to the looser team) and consequently it will give enormous justice to the 'football' if considered as a subject. The **Theory of CESFM** is not to be applicable if the football game ends after 90 min of play with a straightforward result (not draw).

"CESFM": A Proposal for FIFA and IFAB for a new 'Continuous Evaluation Fuzzy Method' of Deciding the WINNER of a Football Match that Would Have Otherwise been Drawn or Tied after 90 min of Play

Recently a new soft computing method called by CESFM has been submitted to IFAB (and FIFA), Zurich, Switzerland on the area of football science and now under consideration of IFAB officials. In this method we consider only those cases of football matches for which the penalty shootout is the existing method for determining the Winner, otherwise that would have been drawn or tied after 90 min of play. The existing method of penalty shootout is having a lot of criticisms due to its huge demerits. In this section we author propose an improved method to FIFA (IFAB) to select the winner by a continuous soft-computing evaluation method. The method is called by **CESFM** which is basically a Fuzzy constructive method. But CESFM in its theory does not exclude penalty kicks by each team as followed presently after 90 min of play. The result of penalty kicks is one of the many components in the method of CESFM. This proposed fuzzy method can easily extract the true dominant team out of the two tied teams using a software called by CESFM-software with the help of continuous real time inputs from the Referee where some of the inputs are given by the Referee using his '**fuzzy pocket machine**' M which is an electronic wireless machine (looking like a mobile phone). It is claimed that if FIFA (IFAB) implements this method of **CESFM** in World Cup Football matches, it will be a true justice to the football world, to the fans, to the players, to the teams (in particular to the looser team) and consequently it will give enormous justice to the 'football' if considered as a subject. The **Theory of CESFM** is not to be applicable if the football game ends after 90 min of play with a straightforward result (not draw).

7.1   Penalty Shootout

The **penalty shootout** is a method of determining a winner in Football matches that would have otherwise been drawn or tied after 90 min of play. A penalty shootout is normally used only in situations where a Winner is needed to be identified and where other methods such as **extra time** and **sudden death** have failed to determine a winner. It avoids the delays involved in staging a **replayed match** in order to produce a result. The term *golden goal* was introduced by FIFA in 1993 along with the rule change because the alternative term, 'sudden death', was perceived to have negative connotations. Following a draw, two fifteen-minute periods of extra time are played. If any team scores a goal during extra time, that team becomes the winner and the game ends at once. The winning goal is known as the 'golden goal'.

If there have been no goals scored after both periods of extra time, a penalty shootout decides the game.

The shoot-out is thus a test of individuals which may be considered inappropriate in a team sport, in particular where a team size is large like 11 (eleven)! Football is a team sport and penalties is not a team, it is the individual. The most important merit of a football game is that the game is played with a complete mutual and continuous understanding, continuous mutual coding/decoding of the 11 players. It is played by a continuous performance of the 11 players together, every second is countable and accountable to them during the 90 min of play. Inferior teams are tempted to play preferably for a scoreless draw or for any n-n draw, calculating that a shoot-out may offer their best hope of victory. Killing play-time as maximum as possible is not an impossible job or tough job of a team during 90 min of play. The 1990 FIFA World Cup was notable for many teams playing defensive football and using time wasting tactics, including even a team of very high caliber who scored only 5 goals but reached the final by winning two shootouts. Hence a common complaint about penalty shootouts is that they only determine the better team in the one, rather narrow, discipline of taking penalty shots, rather than fairly determining the better team in the overall play of 90 min. As a way to decide a football match, shoot-outs have been seen variously as a thrilling climax or as an unsatisfactory cop-out. The result is often seen as an exciting lottery rather than a test of skill. Only a small subset of a footballer's skills is tested by a shoot-out, not the skill of the team.

In this section we propose a new soft-computing method for deciding the Winner of a football match that would have otherwise been drawn or tied after 90 min of play. This method is called by '**CESFM**' which stands for "Continuous Evaluation System for Football Matches".

The proposed theory of CESFM uses the powerful soft-computing mathematical tool "Fuzzy Logic" to compute the winner for such a draw or tied case.

7.2 CESFM

Our proposed method of CESFM (Continuous Evaluation System for Football Matches) is applicable only to such cases of football matches that would have otherwise been drawn or tied after 90 min of play. This method is developed to replace the existing penalty shootout method followed presently by FIFA. In the existing system, a penalty shootout is used to decide somehow a Winner of a football match where the other methods such as extra time and sudden death have failed to determine a winner. The penalty shootout method has a lot of demerits, it does not give satisfaction to the looser. The spirit of the sport is not translated to the result, instead it becomes equivalent to a lottery game in many situations. The probability that the better team will win is not high, because it becomes like that 'the better goalkeeper will win, not necessarily the better team". This is a genuine unsolved problem to the world football fans and teams and players, and thus to the football sport if considered as a subject of study and research.

We propose here a new method CESFM which is based on continuous evaluation scores. The CESFM is a multi-criteria based decision making process. A team's performance, merits and demerits, can be evaluated on the basis of certain significant attributes.

7.2.1 Philosophy Behind CESFM

For a draw/tied case, the main problem is: 'How to choose the better?' It is fact that we must develop a new method such that the better team should win, not the weaker team by any chance.

The "Winner" must be

(i) the overall dominating team (of 90 min), not a suddenly dominating team for few minutes only.

(ii) the overall better team by continuous performance shown by the team, who has contributed more substance of the subject 'football' in today's game by way of quality and performance during the 90 min of play.

The existing penalty shootout method of taking decision on the basis of the results of 5 penalty kicks to find the better or dominating team for declaring the 'Winner' is a weak method as it is a test for one player (Goalkeeper) only against one player of the opponent team (five players, one after another, of the opponent team for five penalty kicks). It does not give justice or satisfaction to the football world or to the football subject.

While a football match goes on, each team plays well revealing its merits in various criteria/attributes continuously at every second. It is difficult to evaluate every second the merit-performance of both the teams. But each team also happens to commit mistakes during play, which are not happening continuously but in discrete moments of time. For example, a team commits one foul after 23 min, another foul after 12 min, then shown yellow card to one player of this team by the referee after 18 min, etc. These events if quantified by good mathematical techniques (using hybrid of soft-computing and hard-computing methods) can surely indicate the amount of discredits, amount of poor performance etc. of the concerned team. These events are well recorded in a database by the organizer for statistical analysis. In our proposed CESFM we input such discrete negative-events in a continuous manner during the 90 min of play, in addition to the positive events. Finally we compute the **Continuous Evaluation System Score** or **CES-Scores** or **CS** (in short) of both the teams comparing which we understand which team is the truly better performer today, i.e. which team is the true dominant team today. The team with higher CES-Score (CS) will be declared as the **Winner**. To compute the CS score of a team we compute CS values of few continuous input parameters corresponding to the team.

7.2.2   Parameters in the "CESFM"

FIFA World Cup is one of the most prestigious events happen on this earth! Almost all the countries in the world enjoy FIFA matches with heart-felt interest. But for the draw cases under consideration in this work here, what are the parameters on which one can decide the **Winner**? Can it be just one parameter (results of 5 penalty kicks at the penalty-shootout at the end) only? Can we ignore the parameters showing continuous performance of each team during 90 min of play? Surely Not!

In our proposed Theory of CESM, there are 16 highly significant parameters as its components for continuous inputs to its software called by CESFM-Software. The CESFM does not exclude penalty-shootout kicks (after 90 min of play) as one of its component inputs. But instead of existing practice of 5 kicks, theory of CESFM introduces 10 kicks with the following **CPS** norms. Let us introduce the notion of **CPS** first of all before we proceed further into the depth of CESFM.

7.2.3   What is Called "CESFM Penalty Shootout" (CPS)?

The 'CESFM Penalty Shootout' (CPS) is different from the existing practice of "penalty Shootout". The 'CESFM Penalty Shootout' (CPS) consists of 10 penalty kicks by each team in the existing manner, with the condition that each player (except goalkeeper) of a team will kick once. None can kick twice. The CPS is one mandatory component in CESFM method in a football match (for the draw cases). Replacement of Goalkeeper is allowed by a team as many times the team wants during CPS. But CPS does not allow replacement of any player (other than Goalkeeper) during the tenure of CPS. Thus the set of 10 players of a team (other than Goalkeeper) becomes fixed for CPS which is exactly the set of 10 players of the team at the moment the 90 min of play be over with draw.

Thus three important norms of **CPS** are the following to be followed during **CPS** play time:

(i)   the 10 players of a team becomes fixed, replacement not allowed,
(ii) replacement of Goalkeeper is allowed by a team as many times the team wants,
(iii) the 10 players will kick 10 penalty shots, none can kick twice. This makes the **CPS** different from the existing practice of "penalty Shootout".

The philosophy behind these three norms of **CPS** is that it becomes a game of 11 players for each team up to certain extent. This philosophy is missing in the existing practice of penalty shootout in the FIFA rules.

7.2.4   What is "Half-Ground of the Opponent" for a Team?

We explain the term "**Half-ground of the Opponent**" for each team X and Y here by a figure. Figure 15 shows that team X is on the left hand side and team Y is on the

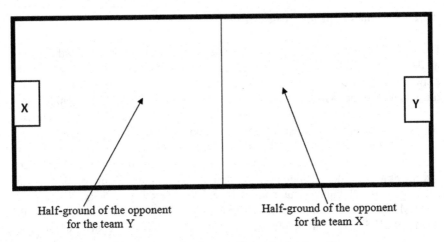

Half-ground of the opponent   Half-ground of the opponent
for the team Y   for the team X

Fig. 15 Half-ground of the opponent for the team X and for the team Y

right hand side during one half of the match. For the team X, the "Half-ground of the Opponent" is shown by arrow mark. Similarly, for the team Y, the "Half-ground of the Opponent" is also shown by arrow mark. These two half-grounds are the partition of the complete ground into two equal halves by the middle half-line.

Thus the "Half-ground of the Opponent for team X" is the half field closer to the goalkeeper of Y, and the "Half-ground of the Opponent for team Y" is the half field closer to the goalkeeper of X.

7.2.5 Positive and Negative Parameters (Corresponding to a Team)

There are 16 parameters (for each team) considered in the theory of CESFM. Few of them are called **negative parameters** and rest of them are to be called **positive parameters**. There are seven negative parameters and nine positive parameters which are listed below. However, more number of decision contributing parameters of continuous nature, positive as well as negative, can be added in the theory of CESFM in future if decided by FIFA or the International Football Association Board (IFAB).

Negative Parameters (corresponding to a team):

   (i)   F1 = Number of Simple Fouls committed (Not shown any card) during the 90 min of play.
   (ii) F2 = Number of Yellow Cards shown by the Referee during the 90 min of play.
   (iii) F3 = Number of Red Cards shown by the Referee during 90 min of play.
   (iv) O = Number of Offsides committed during the 90 min of play
   (v)   H = Number of Handballs during the 90 min of play.

(vi) BPK = Number of Bad Penalty Kicks during **CPS play** (by a 'bad penalty kick' of **CPS** we mean that the kicked ball goes outside the goalpost without touching the goalkeeper or barpost).

(vii) R = Number of Replacement of players made by the Coach (who are not injured) during 90 min of play.

Positive parameters (corresponding to a team):

(i) G = Number of Goals scored (during CPS only).

(ii) BPC = Percentage of Ball Possession at the complete ground during the 90 min of play.

(iii) BPH = Percentage of Ball possession at the "half-ground of the opponent" during 90 min of play.

(iv) SCB = Number of Shots which Collide at Barpost without scoring Goal during 90 min of play.

(v) CK = Number of Corner-kicks (CK) during 90 min play.

(vi) T = Number of Throws during 90 min of play.

(vii) 2G = Consecutive two goals scored (without any goal scored by the opponent in-between) during 90 min of play.

(viii) 3G = Consecutive three goals scored (without any goal scored by the opponent in-between) during 90 min of play.

(ix) nG = Consecutive four or more goals (n > 3) scored (without any goal scored by the opponent in-between) during 90 min of play.

7.2.6 Categories of Fouls

Fouls in a football match are divided into three categories in our Theory of CESFM (Fig. 16):

(i) **Category F1**: Simple Foul (not shown any card)

(ii) **Category F2**: Shown Yellow Card Foul

(iii) **Category F3**: Shown Red Card Foul.

The simple fouls may have various nature and amount of foul-gravity. Similarly the yellow-card fouls may have various nature and amount foul-gravity, and also red-card fouls may have various nature and amount of foul-gravity.

As FIFA recruits the world's best Referees for the football matches, the decision of these Referees are regarded to be the best decisions and hence the final decisions in every respect of their proceedings during play-time of any match. In a match, corresponding to a foul of any of the three categories F1 or F2 or F3, the foul-gravity is a very important but an imprecise ill-defined term which can only be estimated by the Referee by his best possible intellectual judgment on the basis of finite number of significant criteria.

Fig. 16 Three categories of
fouls in a football match

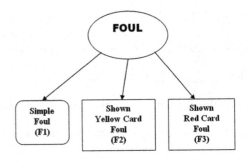

For instance, FIFA could choose the following criteria to estimate the gravity of any foul (F1 or F2 or F3):

(i)   'bad intention mainly for making tactful physical collision',
(ii) 'unfair way of ball possession',
(iii) 'inappropriate body language',
(iv) 'argument with the opponent player(s)', and
(v)   'argument with the Referee'.

The above list of criteria may be revised or improved time to time by FIFA experts (or IFAB experts) and hence the above is not an absolutely fixed list for all time.

7.2.7 CS Value of a Foul (F1 or F2 or F3)

Fouls are negative parameters in the Theory of CESFM as mentioned in earlier. The CS value of a foul (of category F1 or F2 or F3) is always a crisp value. But for this the referee awards a punishment fuzzy set to it at real time of play by his best possible intellectual judgment depending upon the gravity of the foul. This fuzzy set is then de-fuzzified by one of the three algorithms: Algo-1, Algo-2, Algo-3, whichever be applicable depending upon the actual category of foul (as explained in the subsequent subsection) to get the crisp CS value of that foul.

For a foul (which could be of one of the three categories: F1 or F2 or F3), the universal set is the crisp set F given by $F = \{f_1, f_2, f_3, f_4, f_5\}$, where

f_1 = 'bad intention mainly for tactful physical collision',
f_2 = 'unfair way of ball possession',
f_3 = 'inappropriate body language',
f_4 = 'argument with the opponent player(s)', and
f_5 = 'argument with the Referee'.

Note: The elements f_i of the universal set F may be modified, number of elements may be increased/decreased and fixed by the FIFA experts. For the sake of presentation here, we consider the five-membered hypothetical universal set F as mentioned above for the Fouls.

In the Theory of CESFM, for every foul (F1 or F2 or F3) committed by a player (i.e. by a team) team, the Referee takes a fuzzy action against the concerned player (i.e. against the concerned team) by awarding a "punishment fuzzy set" A of the universal set F to the team. The fuzzy set A will be awarded to the team prior to the foul-kick to be kicked by the opponent team at the cost of hardly ten to twelve seconds of the Referee, which is a quite reasonable amount of time.

7.3 "Fuzzy Pocket Machine" M for the Referee

There is a handy small fuzzy pocket machine M for the Referee using which the referee awards a punishment fuzzy set A of F within a maximum of ten to twelve seconds at real time. A fuzzy pocket machine M is a simple electronic wireless machine looking like a mobile phone. Once the Referee awards the 'punishment fuzzy set', the data will be stored immediately in the concerned database in the FIFA-Server for a software called by "CESFM Software" which will be executed if and only if the situation after 90 min is a draw case, otherwise not.

The machine M will have three buttons in the name of F1, F2 and F3 respectively, twelve buttons in the name of: . (dot), 0, 0.1, 0.2, 0.3, 0.4, 0.5, 0.6, 0.7, 0.8, 0.9, and 1 respectively, and two more buttons in the name of 'ENTER' and 'EDIT'. In total there are 17 buttons in the fuzzy pocket machine M (as shown in Fig. 17).

How to input data by the Referee using his machine M?

(i) to input a data like 0.6, the Referee has to press 0.6 button only. To input the data 0, the Referee has to press the 0 button only. To input the data 1, the Referee has to press the 1 button only.

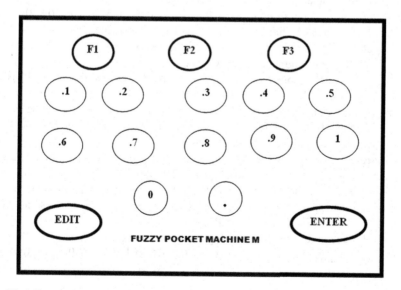

Fig. 17 A Fuzzy pocket machine M for the referee

(ii)   However, to input a data like 0.63, the Referee has to press the following sequence of buttons: . (dot), 0.6, 0.3.

(ii)   Similarly, to input a data like 0.638, the Referee has to press the following sequence of buttons: . (dot), 0.6, 0.3, 0.8.

The architecture of the fuzzy pocket machine M is so designed that on pressing the . (dot) button initially, the each of the other buttons 0.1, 0.2, 0.3, 0.4, 0.5, 0.6, 0.7, 0.8, 0.9 behaves 'without decimal point' till the 'ENTER' button be not pressed. However, the machine M resumes its original status after the 'ENTER' button be pressed. Besides that, a Referee can not input a membership value data which is more than 1 or less than 0 or which has more than three decimal places (fuzzy pocket machine itself does not permit it by its architecture).

The Theory of CESFM being a continuous evaluation method, every amount of credit/discredit of a player cater to the cumulative amount of credit/discredit of the concerned team as a whole. For a committed Foul during play, depending upon the gravity of Foul, the Referee follows the following steps to award a "punishment fuzzy set" A to the concerned player (i.e. to the concerned team) by his best possible intellectual judgment:

Step-1:   Referee press one of the buttons F1 or F2 or F3, and then the button 'ENTER'. (This signifies which of the three categories of fouls is awarded by the Referee.)

Step-2:   Referee inputs one membership value from the closed interval [0, 1] using one or more buttons: . (dot), 0, 0.1, 0.2, 0.3, 0.4, 0.5, 0.6, 0.7, 0.8, 0.9, and 1, and then the button 'ENTER'. (This is the membership value of the element f_1 for the fuzzy set A of the universal set F by best possible intellectual judgment of the Referee.)

Step-3:   Referee inputs one membership value from the closed interval [0, 1] using one or more buttons: . (dot), 0, 0.1, 0.2, 0.3, 0.4, 0.5, 0.6, 0.7, 0.8, 0.9, and 1, and then the button 'ENTER'. (This is the membership value of the element f_2 for the fuzzy set A of the universal set F by best possible intellectual judgment of the Referee.)

Step-4:   Referee inputs one membership value from the closed interval [0, 1] using one or more buttons: . (dot), 0, 0.1, 0.2, 0.3, 0.4, 0.5, 0.6, 0.7, 0.8, 0.9, and 1, and then the button 'ENTER'. (This is the membership value of the element f_3 for the fuzzy set A of the universal set F by best possible intellectual judgment of the Referee.)

Step-5:   Referee inputs one membership value from the closed interval [0, 1] using one or more buttons: . (dot), 0, 0.1, 0.2, 0.3, 0.4, 0.5, 0.6, 0.7, 0.8, 0.9, and 1, and then the button 'ENTER'. (This is the membership value of the element f_4 for the fuzzy set A of the universal set F by best possible intellectual judgment of the Referee.)

Step-6:   Referee inputs one membership value from the closed interval [0, 1] using one or more buttons: . (dot), 0, 0.1, 0.2, 0.3, 0.4, 0.5, 0.6, 0.7, 0.8, 0.9, and

1, and then the button 'ENTER'. (This is the membership value of the element f_5 for the fuzzy set A of the universal set F by best possible intellectual judgment of the Referee.)

Referee can use the EDIT button to change his recently input 'punishment fuzzy set' (membership values) for which the ENTER button is not yet pressed by him, although such cases will be rare by the talent Referees. But after editing if done, he has to press the ENTER button in order to save the modification in the memory of the FIFA-server. Using the fuzzy pocket machine M, the inputs of the Referee are transmitted to the server directly from the playground at real time.

7.4   Three Algorithms Algo-1, Algo-2 and Algo-3 for De-Fuzzification

Corresponding to three categories of fouls (F1, F2, F3), there are three cases and algorithms which are explained below.

Case-1: If the Foul is a Simple Foul

Consider a foul of category F1. The CS value of a foul F1 denoted by CS(F1) will be always one of the numbers 2, 3 and 4. Let A be the 'punishment fuzzy set' awarded by the Referee corresponding to this foul committed by a team. Then the following algorithm called by Algo-1 will be applicable on the basis of α-cut of the fuzzy set A.

Algo-1

If the .5-cut of A (i.e. the set $A_{.5}$) is a null set
      then CS(F1) = 2, Stop. *else*
If the .8-cut of A (i.e. the set $A_{.8}$) is a null set
      then CS(F1) = 3, Stop. *else*
CS(F1) = 4.

Case-2: If the Foul is a Yellow Card Foul

Consider a foul of the category F2. The CS value of a foul F2 denoted by CS(F2) will be always one of the numbers 4, 5 and 6. Let A be the 'punishment fuzzy set' awarded by the Referee corresponding to this foul committed by a team. Then the following algorithm called by Algo-2 will be applicable on the basis of α-cut of the fuzzy set A.

Algo-2

If the .5-cut of A (i.e. the set $A_{.5}$) is a null set
               then $CS(F2) = 4$, Stop. *else*
If the .8-cut of A (i.e. the set $A_{.8}$) is a null set
               then $CS(F2) = 5$, Stop. *else*
$CS(F2) = 6$.

Case-3: If the Foul is a Red Card Foul

Consider a foul of the category F3. The CS value of a foul F3 denoted by $CS(F3)$ will be always one of the numbers 6, 7 and 8. Let A be the 'punishment fuzzy set' awarded by the Referee corresponding to this foul committed by a team. Then the following algorithm called by Algo-3 will be applicable on the basis of α-cut of the fuzzy set A.

Algo-3

If the .5-cut of A (i.e. the set $A_{.5}$) is a null set
               then $CS(F3) = 6$, Stop. *else*
If the .8-cut of A (i.e. the set $A_{.8}$) is a null set
               then $CS(F3) = 7$, Stop. *else*
$CS(F3) = 8$.

7.5   CS Values of Other Parameters in CESFM

Out of sixteen parameters in the theory of CESFM, there are seven number of negative parameters and nine number of positive parameters. The CS values of these 16 parameters play a vital role in the CESFM. The CS values of three negative parameters F1, F2 and F3 are discussed in the preceding subsection. In this subsection we discuss the CS values of all other parameters.

CS Value of negative parameters (of a team)

It is discussed in the preceding subsection that during the 90 min of play, for a Simple Foul F1, the $CS(F1) \in \{2, 3, 4\}$. For a Yellow Card Foul F2, the $CS(F2) \in \{4, 5, 6\}$. And for a Red Card Foul F3, the $CS(F3) \in \{6, 7, 8\}$.

For the other negative parameters, we propose the following CS values in the theory of CESFM (See Subsection 7.2.5).

(i)   For each Offside, CS(O) = 1;

(ii) For each handball, CS(H) = 1;

(iii) For every Bad Penalty Kick, CS(BPK) = 1;

(iv) For Replacement of players during 90 min of play (who are not injured during play), the CS value is given by the following norm:

$$CS(R) = 1, \text{ if 1 or 2 players are Replaced during 90 min} \atop = 2, \text{if 3 or more players Replaced during 90 min.}$$

It is to be noted that the number of replacements of injured players is not considered in the Theory of CESFM for computing any CS value. Also the number of replacement of Goalkeepers during **CPS** play is not considered in the Theory of CESFM for computing any CS value.

CS Value of positive parameters (of a team)

For the positive parameters of a team, we propose the following CS values in the Theory of CESFM (See Subsection 7.2.5).

(i)   For each Goal G scored during **CPS**, CS value CS(G) = 10.
*(It is mentioned that **CPS** is a mandatory component in CESFM method. The goals of the first 90 min of play are not to be considered in CESFM. They are ignored as the situation is a draw/tied case after 90 min of play).*

(ii) If the Ball Possession of a team across the complete ground the during 90 min of play is = x %, then CS(BPC) = x/20.

(iii) If the Ball Possession of a team across the "Half-ground of the Opponent" during 90 min of play is = y %, then CS(BPH) = y/10.

(iv) For each Shot which Collides at Barpost without scoring Goal during 90 min of play, the CS value is CS(SCB) = 5. For each Shot which Collides at Barpost without scoring Goal during **CPS** play, the CS value is CS(SCB) = 1.
(However, if a Shot Collides at Barpost with a Goal during 90 min of play or during **CPS**, then there is no CS(SCB) value for this shot).

(v)   For each corner kick during 90 min of play, the CS value is given as below:
CS(CK) = 1, if the corner-kick is a "**Bad Corner Kick**" (**BCK**) i.e. the ball of that corner-kick goes directly outside the play-ground without touching any player or any location inside the playground or barpost.
CS(CK) = 2, otherwise, for all other cases.
It is to be noted that in case the corner kick hits the barpost, the CS(SCB) value will also be considered as usual, in addition to due value of CS(CK).

(vi) If a team gets a Throw, then for each such throw the CS value is given by CS(Throw) = 1.

(vii) If a team scores two consecutive goals (without any goal scored by the opponent in-between) during 90 min of play, then for each such case the CS value is given by CS(2G) = 1.

This rule is not applicable during **CPS** play.

(viii) If a team scores three consecutive goals (without any goal scored by the opponent in-between) during 90 min of play, then for each such case the CS value is given by CS(3G) = 5 (in such case the CS value of 2G will not be considered).

This rule is not applicable during **CPS** play.

(ix) If a team scores n (n > 3) consecutive goals (without any goal scored by the opponent in-between) during 90 min of play, then for each such case the CS value is given by CS(3G) = 10 (in such case the CS value of 2G or 3G will not be considered).

This rule is not applicable during **CPS** play.

7.6 'CS Score' of a Team

In a football match, suppose that two teams X and Y arrives at a Draw/Tied case after 90 min of play. Therefore this is a situation where the fuzzy method CESFM is applicable. To compute the CS Score of a Team (say, team X), the CESFM-software computes the following values first of all:

(i) Total amount of CS values of all the negative parameters for the team X which is denoted by **NCS(X)**.

(ii) Total amount of CS values of all the positive parameters for the team X which is denoted by **PCS(X)**.

CS Score of the team X is then computed using the formula: CS(X) = 1000 + PCS(X) − NCS(X).

Note: In the above formula for CS(X), there is no significance of the amount 1000. *It is added for no other purpose but just to ensure that the value of CS(X) does not become a negative real number under any circumstances. It is obvious that even if we do not added* 1000 *here, the final result of CESFM-software will not be changed or disturbed.*

Finally the CESFM-software computes the Winner of this football game in the following way.

Who is the Winner by CESFM method?

Suppose that after 90 min of football play, situation is 'draw' case between the two teams X and Y. Then the CESFM-software will be executed in the FIFA-Server to compute the CS score of both the teams X and Y.

If CS(X) > CS(Y) then the team X is the **Winner**. Otherwise, if CS(X) < CS(Y) then the team Y is the **Winner**.

A very rare case is if CS(X) = CS(Y). In that case the **CPS Play** will be repeated once more with 10 penalty kicks by each team, and corresponding data will be sent to the database to run the CESFM-software. If the result yet comes 'draw' then **CPS Play** will be repeated once again and so on, untill the CESFM-software computes the **Winner**. But it is an extremely rare case of very low probability.

7.7 A Hypothetical Example of a Football Game

We explain here with hypothetical data an application of this fuzzy CESFM method in a football game played between two teams X and Y in a FIFA match which has become draw (3-3) after 90 min of play. According to the Theory of CESFM, the FIFA now applies **CPS play** to extract those data which are the last phases of input components for the CESFM-software in this continuous evaluation method. Suppose that after CPS play, the following is the statistics (R-Statistics [10]) for the two teams X and Y as recorded in the database of the FIFA-server corresponding to the 16 parameters of CESFM, shown in Table 1.

The database in FIFA server does also contain the punishment fuzzy sets awarded by the Referee corresponding to the fouls recorded in today's match.

Fouls committed by the team X

The Table-1 above shows that in this FIFA match in total there are five fouls committed by the team X, out of which there are three F1 fouls, two F2 fouls and nil number of F3 fouls.

Table 1 The statistics of 16 parameters for team X and team Y

Serial no.	Parameters	Frequency (X): number of occurrences for team X	Frequency (Y): number of occurrences for team Y
1	F1	3	1
2	F2	2	1
3	F3	0	0
4	O	3	4
5	H	3	2
6	BPK	1	0
7	R	3	1
8	G (CPS only)	4	3
9	BPC	40 %	60 %
10	BPH	26 %	50 %
11	SCB	0	0
12	CK	3 (with one BCK)	5
13	T	6	9
14	2G	0	0
15	3G	0	0
16	nG (n > 3)	0	0

The punishment fuzzy sets corresponding to the three F1 fouls committed by the team X are given by the Referee via his fuzzy pocket machine M which are as below:

X(i) :   the fuzzy set $\{(f_1, 0.85), (f_2, 0.8), (f_3, 0.95), (f_4, 0.85), (f_5, 0.9)\}$, and

X(ii) :   the fuzzy set $\{(f_1, 0.95), (f_2, 0.85), (f_3, 0.9), (f_4, 0.95), (f_5, 1)\}$, and

X(iii) :   the fuzzy set $\{(f_1, 0.85), (f_2, 0.6), (f_3, 0.95), (f_4, 0.85), (f_5, 0.8)\}$.

The punishment fuzzy sets corresponding to the two F2 fouls committed by the team X are given by the Referee via his fuzzy pocket machine M which are as below:

X(iv) :   the fuzzy set $\{(f_1, 0.95), (f_2, 0.9), (f_3, 0.95), (f_4, 1), (f_5, 0.95)\}$, and

X(v) :   the fuzzy set $\{(f_1, 0.85), (f_2, 0.8), (f_3, 0.95), (f_4, 0.9), (f_5, 1)\}$.

While CESM-software executes its Algo-1 and Algo-2, it will get the following de-fuzzified CS values for the team X:

$$CS(X(i)) = 4, CS(X(ii)) = 4, CS(X(iii)) = 4, CS(X(iv)) = 6, CS(X(v)) = 6.$$

There is no F3 fouls committed by any player of team X and hence the Algo-3 of CESFM-software will not be executed for team X for this match.

Fouls committed by the team Y

The Table-1 above also shows that in this FIFA match in total there are two fouls committed by the team Y, out of which there is one F1 foul, one F2 foul and nil number of F3 fouls.

The punishment fuzzy set corresponding to the one F1 fouls committed by the team Y is given by the Referee via his fuzzy pocket machine M which is as below:

Y(i) :   the fuzzy set $\{(f_1, 0.2), (f_2, 0.1), (f_3, 0.15), (f_4, 0), (f_5, 0)\}$.

The punishment fuzzy set corresponding to the one F2 foul committed by the team Y is given by the Referee via his fuzzy pocket machine M which is as below:

Y(ii) :   the fuzzy set $\{(f_1, 0.15), (f_2, 0.2), (f_3, 0.1), (f_4, 0), (f_5, 0)\}$.

While CESM-software executes its Algo-1 and Algo-2, it will get the following de-fuzzified CS values for the team Y:

$$CS(Y(i)) = 2 \text{ and } CS(Y(ii)) = 2.$$

There is no F3 fouls committed by any player of team Y and hence the Algo-3 of CESFM-software will not be executed for team Y for this match.

The CS Values of 16 parameters for each team:

The CESFM-software will thus compute the CS values of all the 16 parameters for both the teams X and Y which are shown in the Table 2.

The CESM-software then computes the **NCS** and **PCS** values of each team as below:

$$\textbf{NCS(X)} = 12 + 12 + 0 + 3 + 3 + 1 + 2 = 33 \text{ and}$$
$$\textbf{PCS(X)} = 40 + 2 + 2.6 + 0 + 5 + 6 + 0 + 0 = 55.6.$$

$$\textbf{NCS(Y)} = 2 + 4 + 0 + 4 + 1 + 0 + 1 = 12 \text{ and}$$
$$\textbf{PCS(Y)} = 30 + 3 + 5 + 0 + 10 + 9 + 0 + 0 = 57.$$

The CESM-software then computes the CS value of each team as below:

$$\textbf{CS(X)} = 1000 + PCS(X) - NCS(X) = \textbf{1022.6} \text{ and}$$
$$\textbf{CS(Y)} = 1000 + PCS(Y) - NCS(Y) = \textbf{1045}.$$

Final output of the CESM-software:

Since $CS(Y) > CS(X)$ in this match, the CESM software outputs that the team Y is the **Winner**, the team X is the looser.

<u>Declaration</u>: The team Y is the '**Winner**' in today's FIFA football match.

Table 2 The CS values of 16 parameters for team X and Y

Serial no.	Parameters	Frequency (X)	CS Value (for X)	Frequency (Y)	CS Value (for Y)
1	F1	3	12	1	2
2	F2	2	12	1	4
3	F3	0	0	0	0
4	O	3	3	4	4
5	H	3	3	1	1
6	BPK	1	1	0	0
7	R	3	2	1	1
8	G (CPS only)	4	40	3	30
9	BPC	40 %	2	60 %	3
10	BPH	26 %	2.6	50 %	5
11	SCB	0	0	0	0
12	CK	3	5	5	10
13	T	6	6	9	9
14	2G	0	0	0	0
15	3G	0	0	0	0
16	nG (n > 3)	0	0	0	0

7.8   Comparing CESFM with Obsolete FIFA Rules

In the above example of football game of 'Team-X versus Team-Y', see that if the fuzzy CESFM method is not applied to this football match between two good teams X and Y, then by the existing FIFA norms of Penalty-shootout method the FIFA will declare the team X as the **Winner (by goals 4-3)** and the team Y as the looser. But any football-fan or expert witnessing this FIFA match will get psychologically shocked with this declaration of Winner, witnessing the 90 min continuous play of both the teams at the stadium and also visualizing the continuous performance of each and every player of both the teams.

This poor decision (of declaring the team X as the winner) is not any fault of FIFA (or IFAB) nor of any of the teams X and Y, nor of the Referee, nor of any football fans, nor of the venue, nor of the host country, nor of the weather-climate or any other reasons. ·

Surely it is due to an <u>obsolete method</u> presently being followed by FIFA (IFAB) on one of the most important issues of football game: "How to select the Winner if the result comes draw after 90 min of play?".

It is due to <u>non-availability</u> of any new innovative modern mathematical (and philosophical) technique which can provide by continuous evaluation of the match performance a good solution and result, retaining the justice to the football, retaining the interest of the game, retaining the interest of the football fans and experts, retaining the interest of the football world as a whole. The author here claims that the soft computing method CESFM is a much improved method for this problem of football subject.

8   Conclusion

This work is on the subject of decoding the 'progress' of decision making process in the Human/Animal cognition systems while evaluating the membership value $\mu(x)$ in a fuzzy set or in an intuitionistic fuzzy set or in any such soft computing set model or in a crisp set, developing a new theory called by "Theory of CIFS".

8.1   Theory of CISF

In the first part of this work, the Theory of CISF is introduced. The two hidden facts about fuzzy set theory (and, about any soft computing set theory) established in this work are:

Fact-1
A decision maker (intelligent agent) can never use or apply 'fuzzy theory' or any soft-computing set theory without intuitionistic fuzzy system.

Fact-2

The Fact-1 does not necessarily require that a fuzzy decision maker (or a crisp ordinary decision maker or a decision maker with any other soft theory models or a decision maker like animal/bird which has brain, etc.) must be aware or knowledgeable about IFS Theory!

It has been philosophically and logically justified that whenever fuzzy theory or any soft computing set theory be applied to any real problem, it happens by the mandatory application of intuitionistic fuzzy system inside the brain (CPU), but it is fact that the decision maker (human being or animal or any living thing having a brain or processor element) need not be aware of IFS Theory! The decision maker is, neither knowingly nor un-knowingly, applying the IFS theory during the progress of any decision making process by any theory; because of the hidden truth that the cognition system has an in-built system software type which spontaneously processes IF philosophy at the kernel.

While a hungry tiger chases a buffalo in a jungle, he decides a lot by his own logic (the logic which is unknown to us). The tiger does not know fuzzy logic or intuitionistic fuzzy logic or type-2 fuzzy logic, etc. But he has his own logic by which he decides and very rightly decides about many issues like:

(i)   which buffalo to chase now (out of thousands buffalos available in his proximity). It is also fact that in many occasions he does not choose the buffalo of his nearest proximity due to some reason, or sometimes he decides to choose to chase a baby buffalo because he does also optimize the chance of his success in the real time scenario.

(ii) even sometimes after chasing a particular buffalo for about 300 m, he decides to give up his run (leading to failure in getting his food in this attempt), or sometimes he decides to shift his target to another buffalo.

All these are real time decision oriented activities done by his best possible judgement, by his own logic which is unknown to us. This tiger may be illiterate according to our rich literature or rich logic, but surely he is literate by his own logic, by his own literature which are unknown to us. Whatever be the logic or literature being practiced by this tiger, the kernel always executes the algorithms of CIFS being initiated by Atanassov Initialization, not by any choice of the tiger. And also this execution is without being dependent upon the knowledge, whatever it be, of the decision maker tiger. The different logic or literature used by different decision makers are analogous to higher level languages operating in the outer annulus sphere (see Figs. 9 and 10) of the cognition system, but the kernel of every decision maker (be it human being, animal/bird, or any living thing which has brain) functions by a common unique machine language of CIFS irrespective of all kind of the knowledge of the decision maker which resides at the outer annulus sphere.

Any decision process by a decision maker for deciding the membership value μ (x) starts with Atanassov's initialization $\langle 0, 0, 1 \rangle$ and then after certain amount of time T (called by Atanassov Processing Time) it converges to the trio

$\langle \mu(x), \vartheta(x), \pi(x) \rangle$ without any further updation of the AT functions. In general in most of the cases of x, $\pi(x) \neq$ NIL. Even if $\pi(x) =$ NIL for one or few elements, it is a very rare situation that $\pi(x)$ will be Nil for all the elements x of the universe X in the IFS proposed by the decision maker. It is justified that it may not be appropriate to use fuzzy theory if the problem under study involves estimation of membership values for <u>large</u> number of elements.

Although fuzzy sets are a special case of intuitionistic fuzzy sets, but the existing concept that "the intuitionistic fuzzy sets are higher order fuzzy sets" is **incorrect** (a similar **incorrect** concept can be imagined if one says that 'fuzzy sets can be viewed as higher order crisp sets'). Rather, the fact is that the IFSs are the most appropriate optimal model for translation of imprecise objects while the fuzzy sets are 'lower order' or 'lower dimensional' intuitionistic fuzzy sets.

Decision making activities of ill-defined problems are a routine work at every moment for every living agent. The same was true during the period prior to the discovery of crisp set theory, even starting from the stone age too. The decision making were better done once the crisp set theory, Venn diagrams, etc. were discovered. After that, a further improvement of decision making process became possible by the discovery of fuzzy set theory. But all these revolutionary tools were very special tools to the decision makers, as these tools suffer from the lack of one or more mandatory and significant dimensions of the architecture of cognition system or brain. Consequently, the lower level processing (processing at the core level of the kernel of cognition system) of the decision making processes could not be decoded or modeled down to human knowledge and literature until the discovery of IFS theory by Atanassov. The architecture of the brain of a living object (human being, animals, etc. whoever be the decision maker here excluding the cases of robots or software which have artificial intelligence) is such that by an external/internal physical attack or damage to the brain the decision process may be made slower in the cognition system, APT may be made very large or even for a patient in coma it could be infinitely large, **but it is fact that under all circumstances the system followed inside the kernel at the lowest level is unique which is CIFS**. The practice or habit of the brain of silently executing CIFS system at the kernel can never be changed by any way. It was fact in stone age, and will continue so for ever. The IFS theory is a very complete and sound, <u>dimensionally optimal</u>. Obviously, while making a fair treatment to any ill defined problem, question does not arise how complex IFS theory be. Question is how to get the best possible results by using the most appropriate and optimal tool. Decoding the 'progress' of decision making process in the human cognition systems (or, in the cognition system of any living animal which has a brain or a processor element, be it of a fuzzy decision maker or an intuitionistic fuzzy decision maker or any intelligent decision maker) while evaluating the membership value $\mu(x)$ to construct a fuzzy set or an IFS or any such soft computing set model, it is observed that the exact algorithm processed (analogous to the execution of machine language program corresponding to any higher level language program) is absolutely nothing but **intuitionistic fuzzy** only, which is not by any choice of the decision maker but by in-built **CIFS**.

However, the intuitionistic fuzzy processing in the cognition system of the membership value μ(x) as a <u>special case</u> many times may converge at **fuzzy** or at the **crisp** output as below for one or more elements of the universe of discourse:

$\left\{\begin{array}{l}\text{(i) from \textbf{Intuitionistic Fuzzy} to \textbf{Fuzzy} to \textbf{Crisp}}, \text{i.e.}\\ \text{(ii) from ``\textbf{equilateral triangular plate } } \Delta \text{ \textbf{of side 1}'' to ``\textbf{line segment} ` } - - - \text{ ' \textbf{of length 1}'' to ``set}\{0,1\}'', \text{i.e.}\\ \text{(iii) from ``\textbf{space}'' } (3-D) \text{ to ``\textbf{line}'' } (2-D) \text{ to ``\textbf{points}'', (only two points).}\end{array}\right.$

The 'Theory of IFS' is purely a choice of the decision maker. The decision maker must be knowledgeable about the 'Theory of IFS' if he wants to use it for solving any ill-defined problem. But the 'Theory of CIFS' is not and never a choice of the decision maker. It is an in-built in the cognition system of every decision maker, irrespective of his any knowledge about intuitionistic fuzzy set theory. Whoever be the decision maker, be it a human or an animal or a bird or any living thing which has brain, the 'Theory of CIFS' is automatically followed and executed inside the kernel of the cognition system.

8.1.1 Difference Between 'Theory of CIFS' and 'Theory of IFS'

The 'Theory of IFS' is a choice of the decision maker. He may choose any of the Soft-Computing Set Theory like: Theory of IFS or Theory of Fuzzy Sets or Theory of Soft Sets or Rough Set Theory or Theory of Type-2 fuzzy sets, etc. depending upon the problem under his study. Thus a decision maker must be knowledgeable about the 'Theory of IFS' if he opts to use it for solving any ill-defined problem. But the 'Theory of CIFS' is not and never a choice of the decision maker. The Atanassov Initialization is not initialized by any choice of the decision maker or by any decision of the decision maker or by any prior information from the kernel of the cognition system to the outer-sense of the decision maker. It is never initialized by the decision maker himself, but it gets automatically initialized. CIFS is a mandatory but silence choice of the cognition system of every decision maker, irrespective of his any knowledge about intuitionistic fuzzy set theory. Whoever be the decision maker, be it a human or an animal or a bird or any living thing which has brain, the 'Theory of CIFS' is automatically followed and executed inside the kernel of the cognition system.

8.1.2 Intuitionistic Fuzzy Set and Interval-Valued Fuzzy Set

An IFS should not be confused with an interval valued fuzzy set. They are two different mathematical models. In an interval valued fuzzy set two membership values (lower value and upper value) are proposed for each element, both considering the 'degree of belongingness', i.e. both are proposed with same objective. But in an IFS, the membership value and the non-membership value of an element are proposed independently, with different objectives. In fuzzy sets (or in interval-valued fuzzy sets) the degree of non-membership is calculated using a mathematical formula, not proposed independently by the decision maker. In IFS the degree of non-membership is not calculated using any mathematical formula, but proposed by the decision maker.

8.2   Pattern Recognition or Object Recognition: An Impossible Task Without CISF

Pattern Recognition or Object Recognition is an one of the earliest and important Decision Making Problems to the earth. We need to recognize new diseases, new stars, new planets, new solar systems, new viruses, new bacteria, etc. to list here a few only out of infinity for the human welfare, ultimately for the welfare of our planet earth. Recently, scientists have recognized the existence of boson but still they are in the process of recognizing whether it is all Higgs bosons or there are other bosons too. This section presents one interesting example of soft-computing solution of a problem of Object Recognition using the theory of IFS, which can not be so well solved if fuzzy theory be applied. The theory of CIFS explains and well justifies that it may not be an appropriate decision to apply fuzzy set theory if the problem under consideration involves the estimation of membership values for a large number of elements.

8.3   Unique Application of Fuzzy Theory in Sports

However by another interesting example it is shown that for a given problem if the decision makers of excellent talent be allowed to give their best possible judgment to the issues (i.e. if they are the best available decision makers on the subject under consideration), then fuzzy set theory will be more appropriate than intuitionistic fuzzy set theory. For this a new theory called by "**Theory of CESFM**" is proposed to FIFA/IFAB and Football-experts of the world with a claim that if FIFA incorporates this fuzzy method in the World Cup Football, there will be a huge improvement in the justice to the game 'football' if considered as a subject. The proposed method is a continuous evaluation method to compute the WINNER using a software known as CESFM-software to be executed at the FIFA server. The fuzzy pocket machine is a simple wireless machine M by which the inputs of the Referee are transmitted to the server directly from the playground at real time of play. The hardware of this machine M can be manufactured by a good company dealing with electronics and communication engineering. The code of CESFM-software can be easily developed by a programmer. CESFM does not exclude the penalty-shootout results. Rather a modified and improved version of the are proposed independently, with different objectives. In fuzzy sets (or in interval-valued fuzzy sets) the degree of non-membership is calculated using a mathematical formula, not proposed independently by the decision maker. In IFS the degree of non-membership is not calculated using any mathematical formula, but proposed by the decision maker.

8.2   Pattern Recognition or Object Recognition: An Impossible Task Without CISF

Pattern Recognition or Object Recognition is an one of the earliest and important Decision Making Problems to the earth. We need to recognize new diseases, new stars, new planets, new solar systems, new viruses, new bacteria, etc. to list here a few only out of infinity for the human welfare, ultimately for the welfare of our planet earth. Recently, scientists have recognized the existence of boson but still they are in the process of recognizing whether it is all Higgs bosons or there are other bosons too. This section presents one interesting example of soft-computing solution of a problem of Object Recognition using the theory of IFS, which can not be so well solved if fuzzy theory be applied. The theory of CIFS explains and well justifies that it may not be an appropriate decision to apply fuzzy set theory if the problem under consideration involves the estimation of membership values for a large number of elements.

8.3   Unique Application of Fuzzy Theory in Sports

However by another interesting example it is shown that for a given problem if the decision makers of excellent talent be allowed to give their best possible judgment to the issues (i.e. if they are the best available decision makers on the subject under consideration), then fuzzy set theory will be more appropriate than intuitionistic fuzzy set theory. For this a new theory called by "**Theory of CESFM**" is proposed to FIFA/IFAB and Football-experts of the world with a claim that if FIFA incorporates this fuzzy method in the World Cup Football, there will be a huge improvement in the justice to the game 'football' if considered as a subject. The proposed method is a continuous evaluation method to compute the WINNER using a software known as CESFM-software to be executed at the FIFA server. The fuzzy pocket machine is a simple wireless machine M by which the inputs of the Referee are transmitted to the server directly from the playground at real time of play. The hardware of this machine M can be manufactured by a good company dealing with electronics and communication engineering. The code of CESFM-software can be easily developed by a programmer. CESFM does not exclude the penalty-shootout results. Rather a modified and improved version of the

However, the intuitionistic fuzzy processing in the cognition system of the membership value μ(x) as a <u>special case</u> many times may converge at **fuzzy** or at the **crisp** output as below for one or more elements of the universe of discourse:

$\Big\{$ (i) from **Intuitionistic Fuzzy** to **Fuzzy** to **Crisp**, i.e.
(ii) from "**equilateral triangular plate Δ of side 1**" to "**line segment** ' – – – ' of length 1" to "set{**0, 1**}", i.e.
(iii) from "**space**" (**3 – D**) to "**line**" (**2 – D**) to "**points**", (only two points).

The 'Theory of IFS' is purely a choice of the decision maker. The decision maker must be knowledgeable about the 'Theory of IFS' if he wants to use it for solving any ill-defined problem. But the 'Theory of CIFS' is not and never a choice of the decision maker. It is an in-built in the cognition system of every decision maker, irrespective of his any knowledge about intuitionistic fuzzy set theory. Whoever be the decision maker, be it a human or an animal or a bird or any living thing which has brain, the 'Theory of CIFS' is automatically followed and executed inside the kernel of the cognition system.

8.1.1 Difference Between 'Theory of CIFS' and 'Theory of IFS'

The 'Theory of IFS' is a choice of the decision maker. He may choose any of the Soft-Computing Set Theory like: Theory of IFS or Theory of Fuzzy Sets or Theory of Soft Sets or Rough Set Theory or Theory of Type-2 fuzzy sets, etc. depending upon the problem under his study. Thus a decision maker must be knowledgeable about the 'Theory of IFS' if he opts to use it for solving any ill-defined problem. But the 'Theory of CIFS' is not and never a choice of the decision maker. The Atanassov Initialization is not initialized by any choice of the decision maker or by any decision of the decision maker or by any prior information from the kernel of the cognition system to the outer-sense of the decision maker. It is never initialized by the decision maker himself, but it gets automatically initialized. CIFS is a mandatory but silence choice of the cognition system of every decision maker, irrespective of his any knowledge about intuitionistic fuzzy set theory. Whoever be the decision maker, be it a human or an animal or a bird or any living thing which has brain, the 'Theory of CIFS' is automatically followed and executed inside the kernel of the cognition system.

8.1.2 Intuitionistic Fuzzy Set and Interval-Valued Fuzzy Set

An IFS should not be confused with an interval valued fuzzy set. They are two different mathematical models. In an interval valued fuzzy set two membership values (lower value and upper value) are proposed for each element, both considering the 'degree of belongingness', i.e. both are proposed with same objective. But in an IFS, the membership value and the non-membership value of an element existing penalty-shootout is introduced called by **CPS** (CESFM Penalty Shootout) which is also a mandatory components in the CESFM method. The **CPS** is a much improved version in the sense of football-interest both meritoriously and philosophically than the existing penalty shootout method of FIFA presently being followed in World Cup Football matches. The proposed method CESFM can do huge justice to the football, can provide huge satisfaction to the football world, for the cases of football matches which have a draw situation after 90 min of play. A hypothetical example is presented to explain how the CESFM computes better results than the results applying the existing 'Penalty Shootout' norms of FIFA. The CESFM is not to be applicable for a decided match in 90 min of play. In the theory of CIFS here it is well justified with several examples that in most of the cases of real life problems Intuitionistic Fuzzy Set Theory of Prof. Atanassov will be the more appropriate tool for applications compared to the Fuzzy Set Theory of Prof. Zadeh. But in the Theory of CESFM the decisions are to be taken by the FIFA Referees of best qualities and of best intellectual capabilities (on the subject) of the world who are expected to have no element of hesitation while proposing membership values. Hence this is a particular case of interest where Fuzzy Set Theory is a better tool for the soft-computing CESFM method compared to the Intuitionistic Fuzzy Set Theory.

9 Future Research Directions

It is seen that for a given element x each of the AT functions $\langle m(x,t), n(x,t), h(x,t) \rangle$ is function of time t. Suppose that $APT(x) = T$. The value $\mu(x)$ comes at the instant $t = T$ from the function $m(x, t)$, as this function $m(x, t)$ finally converges at the value $\mu(x)$ after a course of sufficient growth. The function $m(x, t)$ gets feeding from $h(x, t)$ in a continuous manner starting from time $t = 0$ till time $t = T$. None else feeds $m(x, t)$. The membership value $\mu(x)$ for an element x in a fuzzy set A (or in an intuitionistic fuzzy set or in any soft computing set) can never be derived without the activation of AT functions in the cognition system, and this happens to any brain of human being or animal or of any living thing, irrespective of his education or knowledge. Suppose that the least value of t for which the function $m(x, t)$ attains the value $\mu(x)$ is T_m and the least value of t for which the function $n(x, t)$ attains the value $\vartheta(x)$ is T_n and the least value of t for which the function $h(x, t)$ attains the value $\pi(x)$ is T_h. Therefore $T_m \leq T$, $T_n \leq T$ and $T_h \leq T$.

The following are few open problems:

1. What are other relationships among T_m, T_n, T_h and T?
2. Is it true that $T_m = T_n$ always? If not, then when $T_m > T_n$ and when $T_m < T_n$?
3. Is it true that the curve $m = m(x, t)$ in the mt-plane always becomes parallel to t-axis for some amount of time Δt (>0) just before the Tth instant of time? (similar cases for the curves $n = n(x, t)$ in the nt-plane and $h = h(x, t)$ in the ht-plane?)

4. From the h-bag, in what way the intuitionistic flow happens while filling-up the m-bag and n-bag? Is it sequential but in variable round robin manner?

5. It is seen that $A_m/T + A_n/T + A_h/T = 1$, where each of the three expressions A_m/T, A_n/T and A_h/T is purely a function of the parameter x.

But, $A_m/T \neq \mu(x), A_n/T \neq \vartheta(x)$ and $A_h/T \neq \pi(x)$.

Then what is the significance of the following equality:

$$(A_m/T + A_n/T + A_h/T) = 1 = (\mu(x) + \vartheta(x) + \pi(x))?$$

References

1. Atanassov, K.T.: Intuitionistic fuzzy sets. Fuzzy Sets Syst. **20**, 87–96 (1986)
2. Atanassov, K.T.: More on intuitionistic fuzzy sets. Fuzzy Sets Syst. **33**, 37–45 (1989)
3. Atanassov, K.T.: New operations defined over the intuitionistic fuzzy sets. Fuzzy Sets Syst. **6**, 137–142 (1994)
4. Atanassov, K.T.: Operators over interval valued intuitionistic fuzzy sets. Fuzzy Sets Syst. **64**, 159–174 (1994)
5. Atanassov, K.T.: Intuitionistic Fuzzy Sets: Theory and Applications. Springer, Berlin (1999)
6. Atanassov, K.T.: On Intuitionistic Fuzzy Sets Theory. Springer, Berlin (2012)
7. Atanassov, K.T., Gargov, G.: Interval-valued intuitionistic fuzzy sets. Fuzzy Sets Syst. **31**, 343–349 (1989)
8. Atanassov, K., Pasi, G., Yager, R.R.: Intuitionistic fuzzy interpretations of multi-criteria multi-person and multi-measurement tool decision making. Int. J. Syst. Sci. **36**, 859–868 (2005)
9. Biswas, R.: Decoding the 'Progress' of decision making process in the human/animal cognition systems while evaluating the membership value $\mu(x)$. In: Issues in Intuitionistic Fuzzy Sets and Generalized Nets, vol. 10, pp. 21–53 (2013)
10. Biswas, R.: Introducing soft statistical measures. J. Fuzzy Math. **22**(4), 819–851 (2014)
11. Biswas, R.: CESFM: a proposal to FIFA for a new 'Continuous Evaluation Fuzzy Method' of deciding the WINNER of a football match that would have otherwise been drawn or tied after 90 minutes of play. Am. J. Sports Sci. Med. **3**(1), 1–8 (2015)
12. Bouchon-Meunier, B., Yager, R.R., Zadeh, L.A.: Fuzzy Logic and Soft Computing. World Scientific, Singapore (1995)
13. Dubois, D., Prade, H.: Fuzzy Sets and Systems: Theory and Applications. Academic Press, New York (1990)
14. Dubois, D., Prade, H.: Twofold fuzzy sets and rough sets: some issues in knowledge representation. Fuzzy Sets Syst. **23**, 3–18 (1987)
15. Gau, W.L., Buehrer, D.J.: Vague sets. IEEE Trans. Syst. Man Cybern. **23**(2), 610–614 (1993)
16. Goguen, J.A.: L-fuzzy sets. J. Math. Anal. Appl. **18**, 145–174 (1967)
17. Gorzalzany, M.B.: A method of inference in approximate reasoning based on interval-valued fuzzy sets. Fuzzy Sets Syst. **21**, 1–17 (1987)
18. Kaufmann, A.: Introduction to the Theory of Fuzzy Subsets. Academic Press, New York (1975)
19. Klir, G.K., Yuan, B.: Fuzzy Sets and Fuzzy Logic, Theory and Applications. Prentice Hall, New Jersey (1995)
20. Mizumoto, M., Tanaka, K.: Some properties of fuzzy set of type 2. Inform. Control. **31**, 321–340 (1976)

21. Mololodtsov, D.: Soft set theory-first results. Comp. Math. Appl. **37**(4/5), 19–31 (1999)
22. Novak, V.: Fuzzy Sets and Their Applications, Adam Hilger (1986)
23. Pawlak, Z.: Rough sets. Int. J. Inf. Comp. Sci. **11**, 341–356 (1982)
24. Zadeh, L.A.: Fuzzy sets. Inform. Control. **8**, 338–353 (1965)
25. Zimmermann, H.J.: Fuzzy Set Theory and its Applications. Kluwer Academic Publishers, Boston/Dordrecht/London (1991)
26. http://www.fifa.com
27. http://www.theifab.com
28. http://en.wikipedia.org/wiki/Football_associationc

Abbreviations and Their Significance

CIFS This is the abbreviation for 'Cognitive Intuitionistic Fuzzy System'

Atanassov Trio Functions The three functions m(x, t), n(x, t) and h(x, t) are called Atanassov Trio Functions (AT functions)

Atanassov Initialization The initial values of the AT functions given by: h(x, 0) = 1, with m(x, 0) = 0 and n(x, 0) = 0 is called by 'Atanassov Initialization'

Atanassov Constraint The AT functions are subject to the constraint: h(x, t) + m (x, t) + n(x, t) = 1, at every time t called by Atanassov Constraint

Atanassov Trio Bags These are three imaginary bags: h-bag, m-bag and n-bag

Atanassov Processing Time (APT) After certain amount of time, say after t = T (>0) the processing of the decision making process stops (converges) at the following state, say: h(x, T) = $\pi(x)$, with m(x, T) = $\mu(x)$ and n(x, T) = $\vartheta(x)$. Then we say that for the complete evaluation of the membership value $\mu(x)$ by the concerned decision maker corresponding to the element x, the Atanassov Processing Time (APT) is T unit of time

Atanassov Speed of Convergence The Atanassov Speed of Convergence (ASC) for an element x is a kind of average intuitionistic speed at which its trio functions $\langle m(x, t), n(x, t), h(x, t) \rangle$ converges to the final triplet $\langle m(x, T), n(x, T), h(x, T) \rangle$ i.e. the triplet $\langle (\mu(x), \upsilon(x), \pi(x) \rangle$ where T is the value of APT(x)

Atanassov Absolute Triangle The equilateral triangle PQR of side $\sqrt{2}$ unit in the mnh-space is called the Atanassov Absolute Triangle whose vertices are P (0, 0, 1), Q (0, 1, 0) and R (1, 0, 0)

Decision Trajectory Curve The curve-path PVC of the variable point V lying upon the plane of Atanassov Absolute Triangle is called the Decision Trajectory Curve for x

Atanassov Point of Convergence The end point C = C($\mu(x)$, $\upsilon(x)$, $\pi(x)$) of the curve-path PVC of the variable point V is called the Atanassov Point of Convergence for x

© The Author(s) 2016
R. Biswas, *Is 'Fuzzy Theory' an Appropriate Tool for Large Size Problems?*,
SpringerBriefs in Computational Intelligence, DOI 10.1007/978-3-319-26718-0

Atanassov Line-segment The last decision line-segment L_1ABL_2 is called the Atanassov Line-Segment for the element x

Zadeh Line-segment The base of the Atanassov Absolute Triangle is the side QR of length $\sqrt{2}$ unit called by Zadeh Line Segment, in the name of the great philosopher Prof. Zadeh

Atanassov Triangle The equilateral triangle PAB is called the Atanassov Triangle for the element x whose vertices are the points P $(0, 0, 1)$, A$(0, 1 - \pi(x), \pi(x))$ and B$(1 - \pi(x), 0, \pi(x))$

Intuitionistic μ-index ($i\mu i$) The ratio of the 'membership value amount' to the 'total decided amount' (i.e. excluding the hesitation amount) is called the intuitionistic μ-index (iμi) of x denoted by iμi(x)

Atanassov Indicator (AI) For an element x in the IFS A of the universe X, the Atanassov Indicator (AI) of x is defined by: $AI(x) = \mu(x) \times i\mu i(x)$

FIFA Fédération Internationale de Football Association

IFAB International Football Association Board

CESFM Continuous Evaluation System for Football Matches

Positive/Negative Parameters There are 16 parameters (for each team) in the theory of CESFM. Few of them are called negative parameters and rest of them are called positive parameters

CES-Score (CS) Continuous Evaluation System Score or CES-Scores or CS

Fuzzy Pocket Machine This is a handy small simple electronic wireless machine looking like a mobile phone using which the referee awards a "punishment fuzzy set" A of F within a maximum of ten to twelve seconds at real time

CPS CESFM Penalty Shootout (different from the existing Penalty Shootout)

CESFM-Software Once the Referee awards the punishment fuzzy set, the data will be stored immediately in the concerned database in the FIFA-Server for a software called by "CESFM Software" which will be executed if and only if the situation after 90 min is a draw case, otherwise not

Printed in the United States
By Bookmasters